BRITISH BOARD OF ORDNANCE

SMALL ARMS CONTRACTORS

1689 – 1840

FRONTISPIECE AND COVER ILLUSTRATION

A view of the Tower of London from the east, showing the Pool of London with shipping and London and Southwark Bridges in the background. The Tower Wharf, with the Proof House, Ordnance Workshops and other buildings, the entrance from St. Katherine's, and the east drawbridge are clearly shown. Within the Tower, the Ordnance Office of c. 1770 and the Irish Barracks are visible with the White Tower and the Clock Turret of the Grand Storehouse, which was destroyed by fire in October 1841, behind.

A coloured aquatint *'LONDON PLATE II. Drawn, Engraved and Published by William Daniell, No.9 Cleveland Street, Fitzroy Square, London. August 1st. 1804.'*

William Daniell, (1769–1837), landscape painter and print-maker, specialised in topographical illustrations and paintings and was elected to the Royal Academy in 1822.

© The Board of Trustees of the Armouries (I. 88)

LONDON.
PLATE II.

BRITISH BOARD OF ORDNANCE

SMALL ARMS CONTRACTORS

1689 – 1840

D. W. BAILEY, Ph.D.

PUBLISHED BY
W. S. CURTIS (PUBLISHERS) LIMITED
P. O. Box 493, RHYL, DENBIGHSHIRE,
NORTH WALES, LL18 5XG
GREAT BRITAIN

1999

First published by W. S. Curtis (Publishers) Limited
North Wales, 1999

© All Rights Reserved. No part of this publication may be reproduced, stored in a retrieval system, or transmitted, in any form or by any means, without the prior consent of the Copyright owner. Frontispiece and Cover illustration © The Board of Trustees of the Royal Armouries, Leeds.

British Library Cataloguing-in-Publication Data.

A catalogue record for this book is available from the British Library.

SOFT BACK EDITION ISBN 0 948216 16 6

© The right of Dr. De Witt Bailey to be identified as the author of this work has been asserted in accordance with the Copyright Designs and Patents Act 1988

© 1999. TYPESET IN TIMES NEW ROMAN BY THE PUBLISHER

PRINTED AND BOUND
BY
ANTONY ROWE LTD.,
BUMPER'S FARM, CHIPPENHAM,
WILTSHIRE. SN14 6LH ENGLAND

𝔇𝔢𝔡𝔦𝔠𝔞𝔱𝔢𝔡 𝔱𝔬 𝔱𝔥𝔢 𝔐𝔢𝔪𝔬𝔯𝔶
of the late and greatly lamented

Howard L. Blackmore
1917 — 1999

𝔍n grateful recognition for all that he and his work have done to inspire and encourage my curiosity and research, this study is affectionately dedicated.

When I originally wrote to ask his permission to dedicate this work to him, in giving his cheerful assent Howard wrote that 'I look forward to seeing your effort which, knowing you, will comprehensively tear the arse out of the subject. So all good wishes for its success.'

What higher compliment could a dedicated researcher hope for?

TABLE of CONTENTS

Introduction	Page 11
Chapter 1. An Overview.	Page 15
Chapter 2. Special and Limited Production Arms.	Page 25

Indian Fusils of 1699

Indian Fusils of 1700

Indian Fusils of 1702

Indian Fusils of 1710

Indian Fusils of 1716

Large Fowling Piece of 1737

Lord Crawford's Highland Musket of 1740

Col. Churchill's 10th Dragoon Pistol of 1743

Lord Loudoun's Light Infantry Carbine of 1745

Duke of Cumberland's Dragoon Carbine of 1746

Pattern 1748 Extraordinary Land Service Musket

Pattern 1750 Royal Artillery Officer's Fusil

Indian Fusils of 1753

Marine Officer's Fusils of 1755

Pattern 1781 Burgoyne's Blunderbuss Carbines

Duke of Bouillon's Pistols of 1796

Pattern 1800 Duke of York Hussar's Carbine

Pattern 1801 Life Guards' Rifle

Pattern 1810 Musket-bore Infantry Rifle

Pattern 1812 Life Guards' Rifled Carbine

Indian Guns of 1813

Pattern 1814 Russian Pistols

Part 1:   Page 33

Comprehensive Alphabetical Listing of Contractors.

Part 2:   Page 75

Board of Ordnance Small Arms Components & Production Contractors, listed by the time periods during which they worked, and within each time–span by product, 1688-1840.

1688 – 1715	1739 – 1750	1793 – 1803
1714 – 1717	1754 – 1774	1803 – 1819
1717 – 1725	1775 – 1783	1820 – 1839
1726 – 1738	1784 – 1792	

INTRODUCTION

While there are a number of works which include Board of Ordnance contractors within their listings, to date there has been no work which concentrates upon these men/firms and their contribution to the manufacture of Britain's military small arms. The present work, with its 602 entries in the alphabetical section, is intended to fill this gap.

The importance of these men cannot be overestimated with regard to this manufacture, for at no point in its history has the British Government been able to produce an adequate number of satisfactory military small arms without the aid of the private sector gun trade. The degree of contribution has varied but never has it been possible, despite determined efforts on the part of Government, to wholly eliminate the English gun trade from participating in the production of national small arms.

The initial and terminal dates chosen for this work are based on a number of historical considerations. There was a significant change in emphasis and in manufacturing requirements for military small arms from the time William and Mary commenced their reign in 1689. From their accession, England's military forces played a larger role in European and subsequently world affairs than had ever been the case previously, and the methods by which the Government procured its military arms was gradually modified to meet the new requirements almost continuously from 1689. The introduction of Birmingham contractors to the Ordnance sources of supply began in 1689. Another factor is that there are very few Ordnance weapons identified prior to 1689: there are a few James II (1685-88) muskets, carbines and pistols, but no arms of earlier date known at the time of writing. There will be time enough to expand the coverage backwards when there are sufficient examples of work done by identifiable contractors to merit the work involved. Although 1840 is not the final year in which flintlock military arms were manufactured by Ordnance contractors (that date was probably 1844), all mainstream patterns of arms were by 1840 being produced with the percussion ignition system, and the appointment of George Lovell as Inspector of Small Arms with headquarters at RSAF Enfield, rather than the Tower of London, heralded the recognition not only of new standards and methods of production, but of entirely new designs in Britain's military small arms. With Lovell's appointment the office of Master Furbisher, first held by Richard Wooldridge in the early 18th Century, disappeared and Lovell, primarily an administrator, assumed the duties of the Master Furbisher, most of whose predecessors had been working gunmakers.

The material which has been compiled and presented in this work has concentrated on information which will be of particular use to students and collectors of British military small arms, with a view to providing names and dates for contractors who in most instances will have marked their work in some way, with either their name, initials or a distinguishing mark. While many marks and initials have been associated with known makers, many initials and marks found on arms remain unidentified, and probably belong to workers employed by the contractors too far down the employment hierarchy for their names to have survived in connexion with particular tasks. Regrettably the names of Ordnance Inspectors, who stamped their crowned numbers on most major parts of Ordnance small arms, belong to this category. Other names, such as those of engravers, smallwork suppliers and wooden rammer suppliers have been included to help clarify the division of labour and of components supply which existed throughout the period of this work.

Particularly in the period down to 1722, most of the contractors named in this work, in addition to their production of muskets, carbines and pistols, also carried out large programmes of re-working older arms and components, primarily for use as Sea Service weapons. The manner in which this work was carried out makes it highly unlikely that the finished product will bear the name or other markings of the contractors who performed the work. With the adoption of the Pattern 1718 as the standard for Sea Service patterns for most of the Eighteenth Century, and with the changeover to the Ordnance System of manufacture, this so-called 'dividend work' on Sea Service account was discontinued.

As with much else concerning British Board of Ordnance small arms, the details of contractors who supplied the Board are an on-going study, and updated information will be included either in future editions of this work, or in appendices to larger works in progress.

DEFINITIONS AND ABBREVIATIONS USED IN THE LISTINGS

BIRMINGHAM. As a location the name is used in the broadest sense, especially with lockmakers, to include the surrounding 'Black Country' towns of Staffordshire.

COMPLETE ARMS means different things at different times:

(a) in the period 1688 to 1783 this means exactly what it says, the contractor named supplying all of the components (usually obtained from his fellow specialist gunmakers within the trade). Down to 1764 it would be normal (although not invariable) for the lock to bear the maker's name and sometimes to lack the Crowned Royal Cypher, although bearing some inspection marks and King's Proof on the barrel.

(b) During the 1790s the Board often supplied barrels to the contractors, so that weapons would be supplied '*complete except for barrels.*'

(c) In the period from 1804 to 1840 the term usually means complete '*except for*' certain components which are supplied by the Board of Ordnance to the contractors, the contractors supplying all the components not specifically mentioned. Most commonly this will mean '*complete except for barrels, locks, rammers and* (where appropriate) *bayonets*'.

FURNITURE (Gun Furniture, iron or brass). The mounts of a gun including (from the muzzle downwards) nosecap, ramrod pipes, tailpipe (the lowest rammer pipe), sidepiece (sideplate), trigger guard (called handles by the Ordnance), trigger plate, thumbpiece (escutcheon), butt-trap cover, buttplate.

ROUGH STOCKED and SET UP This indicates that the contractor was furnished from Ordnance Stores with all the necessary components required to produce a complete weapon; the contractor's task was to inlet and shape the stock blank, inlet the barrel and lock, drill the rammer channel and then carry on inletting and finishing the weapon which was then returned to Ordnance Stores as a complete arm although normally without a bayonet, which was fitted by the workers in the Small Gun Office. Arms which have been fabricated by this method will bear a full set of inspection markings on all metal parts and the stock.

SMALLWORK The small ferrous components of a small arm comprising the trigger, swivels, metal and wood screws, bolts/keys, or pins or pinning wire, sometimes rod-retaining springs, sling bar and rings.

THE MONEY SYSTEM:— Pounds, Shillings and Pence

Since money and prices were an integral and critical part of the relationship between the Board of Ordnance and its contractors, it seems reasonable to explain the money in which they dealt. This is the more essential today since even some British readers will have little or no knowledge of the money system of their country for a thousand years until decimalization in 1971.

The basic unit was the Pound Sterling, expressed by the symbol £. There were 20 shillings to the pound, and each shilling was divided into twelve pennies, or pence. Each penny was further divided into four parts called farthings. The silver coins used to represent these sums were as follows:-

Crown = 5 shillings, sometimes known in slang as a dollar because it was the close equivalent to the Spanish milled dollar which was the international currency for most of the period. Four crowns = one pound. The crown was the largest silver coin issued during the period of this work.

Half Crown = two shillings and sixpence = two and a half shillings

Shilling = twelve pence

Sixpence = half a shilling

There were other silver coins such as the Maundy series of 4 pence, 3 pence 2 pence and one penny, some of which passed as currency some of the time, especially the 3 pence, or 'thruppenny bit'.

The copper coins included:—

Penny, minted only in 1797 and 1806

Halfpenny (pronounced hay'pny)

Farthing = one quarter of a penny

During the period of this work prices were generally expressed in one of the following ways, depending on the sums involved:-

£12. 10s. = twelve pounds, ten shillings

£12. 10. 3. = twelve pounds 10 shillings and three pence (pronounced 'thruppence')

£12.10. 3½ = twelve pounds 10 shillings and 'thruppence ha'penny'

£12.10. 3¾ = twelve pounds 10 shillings and thruppence three farthings

Sometimes the pound sign was avoided and the cost expressed entirely as shillings: 40 shillings, usually written 40/- being £2. The cost of the Baker rifle was usually expressed as 97/- instead of £4. 17s. When there was a combination of shillings and pence it was written either as 5s. 10d., or more usually as 5/10, which is the method used in this work.

Occasionally the abbreviation 'gns.' occurs in the records, meaning 'guineas'. These were physically a gold coin introduced early in the reign of Charles II in 1663, originally minted from gold obtained from the Guinea Coast of Africa, hence the name. Although originally worth twenty shillings (i.e. a gold pound) it was stabilized at twenty-one shillings (21/-) from 1717 until it was replaced by the gold sovereign of twenty shillings in 1813. The term survived into the 1980s as a fashionable five percent increase on the pound. Guineas were often expressed as £1. 1s., £5. 5s. &c., as well as 1 gn., 5 gns.

Acknowledgements

My sincere gratitude goes to the staff at the Public Record Office at Kew, who produced, over and over again, the many hundreds of volumes of Board of Ordnance records on which this study is based; and to Erik Goldstein and Stuart C. Mowbray for their valuable assistance.

To my wife Sarah, I owe and gratefully acknowledge an ongoing and unpayable debt of gratitude not only for her continuing support of my work, but her active participation in making many improvements to it.

CONTRACTORS TO THE BOARD OF ORDNANCE
1689 – 1840

CHAPTER 1: AN OVERVIEW

Throughout its long history from pre-Tudor times until its abolition in 1855, the Board of Ordnance supplied ordnance and small arms for Britain's land and sea forces and was itself, from the early 18th Century in charge of military engineers and the Royal Artillery. Throughout its more than three-hundred year history the Board relied primarily on the civilian gun trade of England for the major portion of its military small arms; the secondary portion was provided through foreign purchase from the Low Countries, later Liège. It was the London gun trade which provided the English share of arms, and after its chartering in 1637, the Worshipful Company of Gunmakers controlled the work force which performed virtually all of the required labour in producing the Government's small arms. While England remained largely isolated from European affairs this arrangement caused few problems of real importance, but with the accession of William III and Mary to the throne, 'Dutch William's' dedication to preserving the European Balance of Power against the aggrandizement of Louis XIV created unprecedented demands for small arms with which the current system could not cope. The first changes which were made in 1689 brought the gunworkers of Birmingham and the surrounding 'Black Country' district into the supply network, primarily for barrels, locks and iron gun furniture, but also for complete arms. While this new source of components and arms eased some problems, it only exacerbated others. There was no single design of arms fixed upon to guide more than a single contract, and inspection of the production sequence, apart from proof of barrels, appears to have been minimal and, when present, was carried out by teams of the gunmakers themselves, who were paid extra by the Ordnance to act as examiners of their own and their colleagues' work. More than one hundred gunmaking firms were frequently involved in carrying out an Ordnance contract, with the individual quotas being settled by the hierarchy of the Gunmakers' Company. The result was a hodge-podge of weapons varying wildly in quality of materials and workmanship, and conforming only in very basic terms to a fixed pattern. Deliveries were also irregular, and much delayed or simply non-existent whenever more profitable civilian work was available, as it frequently was. The increasing importance of the Birmingham suppliers was constantly attacked by the London company, even though much of their own materials came from the Midlands. If such a set of conditions may be dignified with the name of 'system', then it might appropriately be called the 'Gunmaker's System,' based as it was with its focal point on the control exercised by the Gunmakers' Company and the Birmingham spokesmen over patterns, prices, workforce and inspection and deliveries. During the period in which this 'system' was operative, 1689 - 1714, the Gunmakers' Company disposed of Ordnance work amongst nearly two hundred separate London gunmakers, virtually all of whom supplied complete arms in addition to carrying out what was known as 'dividend' work, mostly repairing and re-constituting Sea Service small arms. The Ordnance had very little room for manoeuvre.

There was a second set of circumstances which operated very well in conjunction with the above arrangements, and involved the means by which much of the army obtained its small arms. The organization of the British army, since the beginnings of a standing army after the Restoration of Charles II in 1660, has always centred on the individual regiment in its internal structure, training, equipping and attitudes. Supply was one of the chief areas where this applied. The colonel was granted permission by the King to raise a regiment, and from that point onwards the functioning of the resulting unit was very much a proprietary matter. General rules were laid down by various royal orders, and very largely ignored or modified at the discretion of the colonel. Within this framework it is not surprising to find that colonels often designed their own uniforms and arms, and purchased both where they pleased. Those colonels not much interested in the minutia of technical or design detail received their arms from the Board of Ordnance accompanied with agreed responsibility for their care and maintenance. Many others purchased arms from private gunmakers to whatever design was agreed between supplier and customer. Even bore size was not regulated. Until about 1740, a good proportion of the British forces were equipped with these various regimental arms, many of which dated from the first decade of the century. This lack of standardization at regimental level equated well with the situation at the supply level. It is difficult to be precise, but at the conclusion of Britain's participation in the War of the Spanish Succession, perhaps as much as one-third of its army was equipped with non-standard small arms. What followed during succeeding decades was a continuing attempt by the governments of George I and II to rationalize, centralize and standardize both the manufacture and the issue of small arms as well as other areas of military and naval administration. What has been appropriately identified as the 'Ordnance System' was developed in response to the chaos which dominated the supply of arms during the long period of the War of the Spanish Succession under the conditions described above. It was intended not only to bring under direct Ordnance control a larger proportion of the gunmaking processes, but also to centralize the design of small arms, and remove from circulation the large variety of weapons supplied to the troops by regimental colonels, and to create a greater degree of uniformity of both weapons and weapons management, i.e. drill. Whether it was also intended to eliminate the necessity for foreign arms purchases in time of crisis is very doubtful, due to the regularity with which this device was employed with little or no adverse comment in the surviving records.

The System took many years to bring to completion, largely because of the lack of money voted by parsimonious peacetime Parliaments politically committed to the smallest possible armed forces, and to the least possible expenditure on military paraphernalia allowed by the perceived view of national security. The Board of Ordnance appears to have developed the basic lines for the new system during the war years, and to have begun implementing them as soon as was financially practical after the conclusion of peace in 1713. They encountered obvious opposition from the Worshipful Company of Gunmakers of London, who previously had both the design and supply of military small arms largely in their own hands, and indirect obstruction from the opinions, traditions and self-esteem of the regimental colonels and the army bureaucracy whom they were intending to bring under much greater control. They enjoyed the active support of two successive monarchs, George I (1714-1727) and George II (1727-1760) both of whom were blooded soldiers and interested in active military administration. However, the political and constitutional situation in England prevented their support from having the effect the Board might have wished.

The first clear indication of a new system being put into operation came with a contract of 15 September 1714 when the orders were not for complete arms to be supplied, but for '*Land service musket barrels, stocked and sett up with his Maj: Locks, brass work, according to the pattern*'. The barrels, locks and brasswork were taken from supplies accumulated by the Board, and issued to the contractors for mounting in walnut stocks and finishing off. There seems to have been some indecision as to the best pattern for military arms during these years: the 1714 contract called for brass-mounted arms, which were produced during 1715-1717, while the next big contract, for the '*Pattern of the 10,000*' (Pattern 1718) was for iron-mounted arms for both Land and Sea Service. This chopping and changing between each big contract was very much a feature of the earlier Gunmaker's System. The number of contractors executing small arms work during this period dropped from the nearly two hundred of the pre-1714 years to just below ninety, representing a transitional phase in the downward trend. With the completion of the Pattern 1718 arms in 1721 the design of small arms appears to have been concentrated on, and in 1722 a new design, for the first time known as the '*King's Pattern*' was settled upon and promulgated as the standard. Thanks to lack of funds there was a hiatus in production and by the time it was resumed later in the decade the design of 1722 had been dropped in favour of one very similar to that of the Pattern 1718, but with a rounded single-bridle lock in place of the flat, plain lock of the 1718, and with brass furniture (the Pattern 1730). It was also officially laid down in 1722 that colonels wanting arms made for their regiments would, in future, adhere to this basic pattern. This instruction was largely ignored, and it was not until a major re-armament of 1740-41 following the outbreak of the War of Jenkins' Ear (October 1739) that circumstances forced the army into a degree of arms uniformity hitherto unseen in its ranks.

There was a hiatus in small arms production by the Ordnance from 1721 until 1728 when components for a new series of arms began to be produced and stockpiled, and it is really from this time that the fully developed Ordnance System came into operation. All of the various components flowed from the contractors into the bins of Ordnance storerooms in the Tower of London where they were inspected for quality of workmanship and materials, and for conformity to the established pattern. These materials were then issued to selected contractors chosen by Ordnance officials for their reliability in conforming to the patterns and contract time limits, for gradual assembly into complete arms, with the processes sub-divided and inspected at each stage, so that by the time a completed arm arrived back in the hands of the Board's storekeepers, it had been inspected and approved and marked at every stage of production from raw materials to completed arm. It was a highly centralized and carefully controlled system involving a minimum of individual contractors who were obliged to work to a very high standard of pattern keeping. This clearly limited the number of workmen within the unstable gunmaking trade who could be successfully employed. In 1730 when production of the new arms (Pattern 1730 Land Service muskets and pistols) began, eight contractors had supplied barrels (3), locks (5), brasswork (3), smallwork (1) and bayonets (2). For the final stages of production, one firm supplied walnut plank and carried out the rough stocking while two others completed the setting up of the arms. Fortunately for the development of the new System, it was placed under no particular pressure for the first decade of its functioning, and all concerned had opportunities to see how it would work. There were no recorded problems with new production, but subsequent events showed that the quality and uniformity of arms actually in the hands of the troops up to 1740 had been totally unaffected by the new manufacturing system. This situation may be seen as a very clear example of the typical situation in which manufacture of arms was considerably in advance of the issue of new arms to the troops for whom they were being made. There was no officially established 'service life' for a musket during the Eighteenth Century, although the subsequently recognized term of twelve years seems to have been realized through actual experience during this time. When there was no need to re-arm the troops, they were not re-armed, and if regimental company officers were slack in their attention to the state of their men's arms, the arms soon deteriorated to an unserviceable and often dangerous condition. Knowing this situation to exist, and realizing that the efforts of central government to increase its own control over regimental responsibilities were fraught with long delays and little success, the Board of Ordnance exerted as much pressure as it could to keep new arms in their own hands for as long as possible. Only when the arms of a regiment were passed by Ordnance inspectors as being totally unfit for service were new arms issued on specific warrants from the responsible authorities. The wars of Jenkins' Ear and the Austrian Succession (1739-48) put the Ordnance System to its first test of responding to wartime demands, and the evidence suggests that it performed as its designers intended. There was a tremendous drain at the beginning of this period when widespread inspections showed the deplorable state to which the army's small arms had sunk, and a wholesale re-armament took place. For the first time the British army was equipped with a uniform pattern of musket. The only area in which significant expansion of the System was required was that of the setters up, since the components makers were apparently able to furnish more materials than the existing setters up could complete in a reasonable time. Hence, between July 1742 and September 1743 five new setters up were added to the two regulars, with three additional firms being added in March and June 1746 and June 1747 for the duration of the war. The rough stocking continued to be done by a single firm, with brass furniture and smallwork also emanating each from a single source. Barrels and locks came from four contractors, with locks only from two others. Several thousand complete arms were supplied between 1742 and 1746 by one partnership.

Wartime experience suggested that the expensive and heavy iron rammer which had been tried as early as 1724 in British service, was indeed a highly desirable improvement in the efficient use of the infantry musket, and with the return of peace in 1748 the Board entered on a programme of converting the arms in the hands of the troops to take the new stronger rammer. The first conversions were carried out by regimental armourers still with the forces in Flanders from the Spring of 1748 and in 1749, probably amounting to about 20,000 muskets being converted. But after 1749 virtually all of this work was carried out by the workforce of the Small Gun Office in the Tower of London, with one contractor supplying the iron rammers, tailpipe springs and brass for sleeving pipes. There was almost no other contractor activity between the end of the war in 1748 and the build-up for the next conflict in 1755, with the exception of several thousand new-production muskets fitted with iron rammers (Pattern 1748) which were set up by nine contractors between December 1748 and March 1749. The System thus proved its elasticity without jeopardizing the tasks undertaken by the Board.

At the opening of the Seven Years' War the Board experienced its first recorded clash with the London gun trade since the founding years of the System, when a combination of contractors refused to work unless prices were considerably increased over what they had been during the previous war. Thanks to the business acumen of a single contractor, John Hirst, the Board overcame this problem, and Hirst became virtually the only setter up employed by the Board for almost twenty years. He was able to command a sufficient workforce to perform the work required by the Board, completing as many as 60,000 muskets in a single year, and during the Seven Years' War the Ordnance System unquestionably achieved its greatest results with the smallest number of problems. The comparatively recent re-armament of the army, and the relatively small percentage of the troops actively engaged during the previous war meant that the stores were full of arms of a modern design and in good condition at the outbreak of the new war, and there were fewer foreign arms purchased during this war than at any other time. Among other benefits, this allowed increased attention to the design and production of a series of new arms for various branches of the services which had either been neglected in the past, or had recently come into being to meet newly perceived military requirements. Thus there were muskets for the newly raised marines and militia, and carbines for the cavalry, light dragoons, Highland regiments, artillery (officers, men and cadets), and various elite units. The rifle made its first combat appearance in the hands of British troops, although it was supplied from abroad and apparently used by troops who were largely of foreign birth. With the proliferation of new designs produced, and a total of nearly 305,000 arms delivered during the period by forty-seven contractors, the Ordnance System clearly reached its apogee during the Seven Years' War. Britain's greater involvement in the war had produced much higher casualties amongst both troops and their arms than during the previous conflict, and production of small arms continued at a reduced level for several years following the peace of 1763. Also in the wind was a major re-designing of the infantry musket to allow greater mobility, facility of use, and comfort for the troops. The acceptance of this new design, the Short Land Pattern, in 1768, produced a spasm of components manufacture into the early 1770s, and most of these parts were turned into complete arms before the American Rebellion erupted in 1775. There was, in fact, a pause in production of the new short musket between July 1772 and March 1775. At the same time that the Short Land Pattern became the standard line infantry arm, new carbines were produced for the serjeants of grenadier regiments, dragoons and artillery, introducing a greater degree of uniformity in design as between the various types.

That the Ordnance System failed to achieve the objectives apparently set for it by the Board during the American War for Independence was probably due to a number of factors not directly attributable to the structure or operation of the System itself. It seems likely that the Board over-reacted to the crisis of French entry into the war in 1778, and demanded more than the available workforce could produce without an unacceptable lowering of quality. Even given the raising of the Volunteers from 1779 which created a new and large demand for arms, the ultimate production achieved by various methods was very far above any possible needs of the service. In addition to this domestic over-production, more than 100,000 foreign arms were imported during the war. It was an early instance of over-kill, and one in which the resources of the domestic gun trade were apparently unable to meet the demands of Government if confined to the rigorous standards previously imposed. The alternative was a partial breakdown in the System, and a return to the earlier practices where contractors were allowed to turn in complete arms which had not undergone the several individual stages of inspection before the final acceptance inspection. This would allow the utilization of a portion of the workforce which was normally considered as unable to work to Ordnance standards. Although this was mainly confined to Sea Service muskets, there was also a large number of Land Service muskets accepted on this older system, some with components varying from the established King's Pattern (notably the sideplate). There were also more labour troubles, this time with the Birmingham lockmakers, an area where the Board had almost no alternative source and where no individual saviour appeared to rescue them as had John Hirst at the beginning of the previous war. The result was a delay of almost two years in the significant production of arms for lack of a major component. This was probably a relevant factor in the subsequent over-production, coinciding as it did with the entry of France into the conflict and turning it into a virtually global commitment for Britain's armed forces. If, however, the end justified the means, then the Board certainly did achieve its objective in the total acquisition of small arms. The Ordnance System itself produced over 298,000 arms using forty-eight contractors (compared with some 305,000 from forty-seven during the previous war), but the Board acquired a total of about 504,000 arms through the use of other domestic and foreign sources. The number of arms obtained by the use of the established System compare favourably with the commitments of the armed forces, remembering that Britain's involvement during the Seven Years' War included sizeable bodies of troops in North America and on the Continent, while the American War was, so far as troop concentrations were concerned, largely a matter of troops in North America and the Caribbean, and a mixed anti-invasion force at home. It seems almost as if the Board lost control of its manufacturing agencies, and let the supplies run on almost unheeded.

The conclusion of the American War and the war with the Continental powers in 1783 found the Ordnance stores full to overflowing, and the basis of its long established System badly undermined by wartime expediencies. There followed a decade during which there was very little contractor production, and that by sixteen contractors and confined to secondary arms such as various carbines and a few Sea Service muskets. Several large auctions at the Tower disposed of tens of thousands of muskets by mid-decade. Component production was minimal. The personality and views of the Master General who assumed office in 1782, the third Duke of Richmond, ensured that the vacuum thus created in Ordnance affairs would be filled with a programme of modernization, experimentation and general improvements in the quality of design and materials. This was the era of the Hennem screwless lock, the Egg-Crespi breech loading carbines, and the many experimental arms of Henry Nock which culminated in the designs for the Duke of Richmond's muskets. But in all of this small-scale work there was no requirement for the Ordnance System to be used, and it was apparently allowed to disintegrate from disuse. Simplification of arms design, improvement in the quality of materials, and lack of need for large quantities of arms seemed to suggest to those in power that there would be no real need for the elasticity of an Ordnance System when the next emergency arose. Inspection procedures had been pruned during the war, and the processes of rough stocking and setting up had been allowed to be performed by the same firms, and this was continued after the war. There would be no return to the older methods. During the decade following the end of the war many new ideas in small arms design were tried, and most were rejected on the grounds of cost vs. actual utility in service, but by 1790 it had been decided to change the basic design of the infantry musket. This plan appears to have been thwarted only by the outbreak of the French Revolutionary Wars in 1793, by which time the tooling up and new gauges for the radically different Duke of Richmond's muskets were not well forward in preparation, and there had been too little opportunity for the workmen to familiarize themselves with the new requirements, a factor still vital to efficient performance in a pre-Industrial Revolution craft industry.

With the outbreak of war all thoughts of attempting to return to the old Ordnance System went by the board in the face of overwhelming demand. The number of contractors rose steadily to a total of seventy-eight by the Peace of Amiens in 1802. Not only did the regular army require its arms, but the county Militia and a Volunteer force were called into being, both of which required far larger quantities of arms than those needed by the army alone; in addition there were thousands of refugees from the French Revolution who were organized into regiments to fight the new political force, and who were to be armed by the British Government. In the face of this unprecedented demand Liège was lost through French occupation as a basic source of supply. British purchasing agents combed the European markets to acquire whatever they could find of serviceable arms. The supplies of military pattern arms in the hands of the domestic trade were also absorbed. Finally, in 1797, the Board accepted that a lowering of quality and simplification of design for the basic Land Service musket would have to be officially sanctioned in order to make use of those portions of the workforce who could not produce to the usual Ordnance standards. Arms of the 'India Pattern' had been initially purchased directly from the East India Company in 1793, and by 1795 they were beginning to be produced directly for the Ordnance, but in 1797 the adoption of the cheaper design was made official and a large-scale production programme initiated.

The Ordnance never abandoned the intention of returning at the first opportunity to a high quality Land Service musket for the infantry, and with the Peace of Amiens in 1802 an attempt was made to begin production of the New Land Pattern series. This was abruptly curtailed by the renewal of war in 1804, and the India Pattern continued as the standard arm in the hands of most regiments for the remainder of the flintlock era which ended in the early 1840s. Maximum utilization of the entire domestic gun trade came closest to achievement in the years from 1804 to 1815, when the Tower of London itself became, for the first time, a manufacturing facility on a full-time basis (the Royal Manufactory of Small Arms), and the totality of the Birmingham trade were either brought into the Board's manufacturing organization, or else prevented from operating outside of it. Proof and inspection facilities were established at Birmingham (confusingly called 'the Tower'), and complete arms were constructed by the contractors there, as well as components, with the London trade continuing to act chiefly as setters up, or, as they were now termed, rough stockers and setters up. By 1815 some two-and-a-half million arms had been produced, more than two million of which were India Pattern muskets, by 164 contractors, only about twenty less than had comprised the contractor list prior to 1714.

Can the Ordnance System, then, be credited with performing successfully the role established for it by its originators in the second decade of the Eighteenth Century? The answer to this broad question is a qualified 'Yes'. The System worked without notable fault during the wars up to 1778, when Britain's military and naval commitments were expanded beyond what they had ever been before, and certainly beyond what could have been envisaged in the early part of the century. It may be said that times changed and the System did not change with it, but this is not strictly the case. Times changed, but the overall number of skilled gunworkers who formed the workforce utilized by the contractors, did not increase significantly. The bulk of Britain's arms manufacture was in the middling and lower categories of small arms – cheap quality sporting arms and handguns, and weapons for trade with non-European countries – which did not require working closely to a pattern or the use of top-quality raw materials. This requirement did not change until the actual disappearance of the 'trade' category of arms well into the Twentieth Century, while the military requirements of Government did both change and expand. The basic quality guidelines of the Ordnance System did not change, but as quantity requirements increased the numbers of qualified workforce did not increase to meet this new demand, since the demand was extremely uncertain and periodic, and it was not considered worthwhile to train men to meet it. Therefore, while the times and military requirements changed, one essential part of the Ordnance System, the skilled workman, did not grow in number in line with the increased requirements. It is in this way alone that the Ordnance System principally failed to meet it goals, and this was a factor outside the control of the Ordnance to remedy without the establishment of their own gunmaking facilities such as came into being at Enfield after 1815. Indeed, the civilian gun trade had always been, and would continue to be, a most unstable industry, with its workforce expanding and declining as the demand dictated. It was for this reason that, finally, in 1804, the Government took the first tentative steps towards establishing a permanent Government-based gun making facility, initially based on the old Armoury Mills complex in Lewisham, Kent, which commenced operations in 1808, and which by 1818 became the Royal Manufactory of Small Arms at Enfield Lock in Middlesex. It was hoped that this facility would provide a permanent cadre of highly skilled workmen who could swiftly train up the required additional workforce in emergency circumstances.

During the century from 1714 to 1815 the Board of Ordnance, having established a system of manufacture which, given the constraints imposed by the financial limitations originating in annual Parliamentary grants, managed to gain and maintain sufficient control over the civilian gun trade which had to form its focal point, to enable it to produce a wide variety of small arms for all of the armed forces to a higher-than-normal standard of workmanship and finish, and (from 1755) to closely gauged patterns for all the components. As it was understood at the time, in a pre-Industrial Revolution non-factory complex, a high degree of standardisation was achieved in the manufacture of the mainstream weapon – the infantry musket – and in certain major components for other basic arms such as the barrels and locks for carbines and pistols.

Production of the India Pattern musket ceased in October 1815. The New Land Pattern, in Infantry and Light Infantry patterns, had been produced in reduced numbers since mid-1811, and production was concentrated on these arms from mid-1814 when it was thought the war was over. Production of both types ground to a halt by mid-1817. In addition to these infantry arms, just under 500 musket-bore Heavy Dragoon carbines were delivered by the contractors between October 1816 and January 1817.

The goal of marginalizing the contractors as far as possible seems to have been behind the gradual build-up of a Government-owned and operated small arms manufacturing complex since at least 1804. With the end of large-scale European war in 1815 movement toward this goal was accelerated, and by the time the various contractors completed their work on the New Land Pattern muskets in mid-1817 and a few additional cadet muskets for the expanding military academy in 1818-19, it looked as though the new establishment at Enfield would be in clear control of future military small arms manufacture for the British Government. Such did not prove to be the case, and indeed the contractors were never to be downgraded to a less than vital position until the large-scale introduction of machinery to the manufacturing process in the mid to late 1850s. The first renewal of production after the gap described above, came in 1820 with a delivery of 450 Infantry Rifles by eleven London contractors. Since there was no subsidiary components production it may be assumed that either the components were produced at RSAF Enfield (for which there are no detailed records) or, much more likely, that the required parts came from the Ordnance Stores in the Tower. During the following year some 300 pairs of Sea Service pistols were issued from Store and were cut-down from a 12-inch to a 9-inch barrel by six London contractors, with a tailpipe and spring being fitted, and a steel rammer supplied. This was almost certainly done to establish a price for this type of work, which was then carried out at Enfield, since the number of pistols of this description which have survived far exceeds this small number. During the same year, ten London contractors produced some 1,700 pairs of Land Service pistols, with the standard 9-inch carbine-bore barrel and swivel rammer, and these may also have been a pricing experiment to establish what this type of work cost in the trade as opposed to what it could be done for at Enfield.

The production of the Pattern 1823 Infantry Rifle between 1823 and 1827 was a joint effort between the private trade contractors and RSAF Enfield, each being intended to produce half of the order of ten thousand rifles. The majority of the contractors were those who had provided components and complete arms during the late war, or their successors. This was the last large order in which the contractors participated significantly during the flintlock period.

During the mid-1820s several contractors also supplied components which were probably intended for use by Enfield in constructing or repairing arms. These consisted of such materials as sidenails, rammers and ribs for musket-bore Heavy Dragoon carbines, to be followed in early 1828 by orders for locks for both carbines and pistols of this description from nine contractors.

Contractor Numbers during the several production periods 1689 – 1840. These include firms which worked in more than one period, and also those which changed their firm styles.

1689 – 1714	183	1756 – 1774	47	1793 – 1802	78
1715 – 1721	86	1775 – 1783	48	1803 – 1819	164
1730 – 1750	32	1784 – 1792	16	Post 1819	120

AN OVERVIEW

The alphabetical list in Part I contains the names of contractors who supplied arms and / or major components of arms to the Board of Ordnance during the period from the accession of William III and Mary to the effective end of the production of flintlock arms by the Ordnance. Examination of the list will show that the contractors fall into several clearly defined periods, generally coinciding with wartime years and their aftermath, and these are grouped together as appropriate in Part II.

Whom amongst the many names listed below can be considered as the 'most important'? Speaking in terms of quantity in their several areas, there is no doubt that Waller, as the primary rough stocker; Hollier as the supplier of furniture; Barber and Pickfatt as the two setters up, dominate the 1730-42 period. With Waller and Hollier continuing, a small group of setters up joined the force 'for the war' from 1742-9, of whom the most important was Birkell. From 1756, with Waller still dominating the rough stocking, the Loxhams replace Hollier for bayonets and Hartwell & Mayor replace him for furniture, Hirst becomes virtual monopoly setter up until 1763, with Farmer, Jordan, Oughton, and Grice, as the leading components suppliers. These firms remain unchanged until, in 1769, Loder joins Waller in rough stocking. From 1777 Hirst's position is eroded principally by John Pratt, and by a small number of London setters up who supply, chiefly, complete Sea Service muskets. During the inter-war period Jonathan Hennem and Henry Nock carried out most of the very limited production of Ordnance small arms. With the outbreak of a Europe-wide war in 1793, the Ordnance was not well prepared to meet the demands created, not only by an expansion of its own forces, including several levels of militia and volunteers, but by refugee forces from conquered areas of Europe who were taken into British pay. During the early years of this period the pre-existing facilities of firms like Thomas Barnett, Durs Egg, Harrison & Thompson, Memory & Wright and Henry Nock coped as best they could, and after 1797 the load was carried chiefly by David Blair, Michael Brander, Galton & Son, Thomas Gill, Ketland & Walker, John Whately and Robert Wheeler. Note that there was a shift in the location of the most significant suppliers during the decade from London to Birmingham. During the final period of significant Ordnance small arms production covered by this work, 1804–1815, the pre-eminent area of production remained in Birmingham and the surrounding Black Country, with the London facilities being augmented by the addition, in 1804, of the Royal Manufactory of Small Arms in the Tower of London. This new centre was, however, dependent upon the Birmingham and London contractors for component parts such as barrels, locks, furniture and rough stocks, which were assembled into complete arms by the much expanded Tower workforce. The Birmingham giants – Ketland, Walker & Co., Ketland & Allport, R. & R. Sutherland and Samuel Galton – produced some 705,000 of the nearly two million India Pattern muskets made during this period by the contractors, the top four London producers – Brander / Brander & Potts, James Thompson & Son, Thomas Ashton and Thomas Barnett achieving only 176,600 of the total. Each of the Birmingham makers produced in the six-figure range while the London makers each produced in the 40,000 range. During this final expanded period of production the contractors produced a total of twenty-five patterns of arms, plus at least five identifiable patterns of Extra Service muskets.

CHAPTER 2

SPECIAL and LIMITED PRODUCTION ARMS.

Throughout almost all of the period covered by this work, it was broadly Ordnance policy to allow individual contractors who were gunmakers in their own right to supply small numbers or one-off orders of specialist arms whose design and production would have disrupted mainstream (i.e. musket) small arm production. The presumption was that at least they would be approved beforehand and inspected by a competent Ordnance inspector before being received and issued, thus avoiding the payment for and failures of completely unsatisfactory weapons; most were also subjected to the King's Proof as a further safeguard. The flexibility of the contractor system allowed the Board of Ordnance to supply particular wants for various military, naval and colonial purposes, always provided that there was approval from the relevant higher officials, usually a Secretary of State. It was part of the 'systematising' or perhaps 'centralizing' mentality to accept that it was better to carry out these specialist orders within the gunmaking structure of the Board than to let contractors in the private sector run rampant as to quality and price.

The Indian Fusils of 1699

Apart from their early date, the most interesting feature of this group is that they were iron-mounted, perhaps not the first but certainly the last Indian guns to be so fitted for the Board of Ordnance. The order was for 180 guns with 3' 10" fine-bored round barrels, dark walnut stocks fitted with round bridle locks and with *'hollowed iron furniture'*, i.e., wrought iron, and with King William III's Crown and Cypher (initials W.R.) engraved on the lockplate and escutcheon (thumbpiece). They were supplied complete by five of the London gunmakers, regular contractors to the Board: Thomas Austin, Henry Bancks, Henry Crips, Robert Silke and Godfrey Taylor.

The Indian Fusils of 1700

Only a few months after the completion of the iron-mounted Indian gift fusils, in June 1700, a second order for another 400 fusils, this time brass-mounted, was distributed amongst selected London contractors. These fusils were to have 3' 10" barrels, with stained and varnished beech stocks, with brass heelplates [buttplates], sideplates and thumbplates, with brass swaged pipes, and round locks (no mention of bridles) engraved with the King's Cypher. These were, at 18/- considerably more costly than the previous order. The stocks are now of beech rather than walnut, and the King's Cypher is to appear only on the lockplate, despite the stock having a thumbpiece. These arms were provided by Thomas Austin, Henry Bancks, Henry Crips, Edward Nicholson, Robert Silke and Godfrey Taylor.

The Indian Fusils of 1702

A little over two years after the previous order for Indian guns was completed another order, this time described only as *'Jamaica Fashion'* fusils *'according to the Pattern sent to the Board'*, was distributed amongst six London gunmakers, once again including the Furbisher, Henry Crips, as well as Thomas Austin, William Moore, Edward Nicholson, James Peddell and John Sibley. The Jamaica Fashion most probably refers to two features both of which are subsequently identified with this type of gun: a ketch-lock (i.e. dog-lock) and a butt design with a deeply curving underside ('roach' or simply 'fish-bellied') and a high comb well undercut where it joins the grip- in other words like a 'Buccaneer' gun of the period.

The Indian Fusils of 1710

The next order for Indian gift fusils involved the same total number, 400, and the same cost, 18/- each, but was divided amongst twenty-five of the London contractors (noted under their names in Part I). These guns again had 3' 10" barrels, but the stocks (wood unspecified, but almost certainly beech) were to be *'painted and spotted according to Pattern'*. This is the first mention of what came to be a common feature on Indian guns – the spotted stocks. Did the Ordnance copy from an existing example, or did someone involved in this contract initiate the feature? It is also of interest that while commercially supplied Indian guns continued for decades after this time to be fitted with either 54-inch or 48-inch barrels, the guns made for the Ordnance had 46-inch barrels in line with the standard musket length barrel of the period.

The Indian Fusils of 1716

The final order for Indian gift guns before a long gap in Ordnance production of such arms was placed in June 1716, and completed the following month by eleven London contractors (noted under their names in Part I). Firmly in line with the newly established 'Ordnance System' of manufacture, these 400 fusils were to be *'stocked and set up with His Majesties Barrels, Locks and Brasswork according to pattern'* at 6/6 each, reflecting the growing influence of the new Ordnance System of manufacture. There are regrettably no further design details. This group got off to a bad start: having been loaded aboard the *'Antelope'*, the ship was driven ashore at the mouth of the River Thames and wrecked. Fortunately, divers recovered all of the guns and they were returned to the Tower. By a warrant of 23 November 1716, they were issued to four contractors, Edward Loader and three of the original contractors, George Aldridge, Robert Looker and Richard Sinckler, who *'cleaned, repaired, cleansed and varnished according to the contract'* for 2/- each.

The Large Fowling Pieces of 1737 – 1738

Eight of these pieces were fabricated, presumably in connexion with the colonization of Georgia then taking place. So far as can be determined from the one known surviving example (which was professionally altered to percussion, half-stocked and fitted with a rib under the barrel in the 19th Century) it was simply an enlarged musket of the current pattern but with a longer and heavier-walled swamped barrel not equipped to take a bayonet and fitted with a steel blade foresight. It was approximately the same length as the Long Land Musket but shorter and lighter than the Wall Piece.

Thomas Hollier supplied eight sets of brass furniture (weighing in total seventeen pounds) in December 1737. In design the brasswork is identical to the current Land Service furniture, only larger. The overall length of the trigger guard is, for instance, 11⅝", and of the buttplate tang 6⅜", the maximum width of the latter being 2½". This is the first arm to use the 'new' pattern trigger guard design, with the 'hazelnut' front finial, which first becomes standard on pistols in 1738 and then on Land Service muskets from 1742.

Richard Waller rough stocked the eight *'long fowling pieces'* for 10/- each, in September 1737.

Edward Jordan supplied the barrels (to judge by the E I markings on the breech) and set the eight pieces up. Although of a similar length to the Land Musket barrel at 46⅛", the barrel itself is much larger in all its other dimensions, being .79 calibre with ¼" thick barrel walls at the muzzle, and measuring 1⅝" across the breech and 1³/₁₆" across the muzzle, the swamping effect creating a measurement of 1⅛" at its narrowest point eleven inches from the muzzle.

The large locks (lockplate measures 7⅞" x 1½") may have been supplied by the Small Gun Office workers since there are no contractors' markings internally, only a crowned number inspector's stamp and the initial **R**. External engraving on the surviving example has been erased.

Lord Crawford's (43ʳᵈ later 42ⁿᵈ) Highland Musket of 1740

In May 1740, Lewis Barbar delivered into Ordnance Stores 229 muskets with bayonets *'to the Pattern of Lord Crawford's Regiment'* at a similar cost to the current Land Service musket, £1.10.0. There were no further recorded deliveries of this pattern of musket. No example of what is very probably a 'light' musket has been identified at the time of writing.

Colonel Churchill's (10ᵗʰ) Dragoon Pistols of 1743 — 1745

An armament of specially designed pistols for Charles Churchill's 10ᵗʰ Dragoons was provided initially by the Master Furbisher, Richard Wooldridge, who delivered 330 pairs in March 1743, and a second delivery of 66 pairs in November of the same year, both priced at 35/- a pair. A final delivery of 84 pairs in August 1745 by James Barber priced at 36/- per pair completed production. This is the second in a series of 'light arms' produced for various British regiments during the War of the Austrian Succession, when the concept of 'light' introduced by exposure to Austrian light cavalry, began to influence British military thinking. With ten-inch (as opposed to the conventional twelve-inch) barrels in pistol-bore, i.e., ·56" calibre rather than the increasingly favoured carbine-bore (·66") for pistols, and with lighter brass furniture of a simplified design, these pistols characterize the move towards smaller and lighter weapons for more mobile troops. These are also the first of a number of ten-inch barrelled pistols to be used in the British service. At 35/- and 36/- a pair they were considerably more expensive than the current Land Service pistols at 26/- a pair.

Lord Loudoun's Light Infantry Carbine 'Short Musket of Carbine-bore' of 1745

The next in the Barber series is a 39-inch barrelled carbine-bore light infantry carbine – the first 39-inch barrel in British service – of which James Barber delivered 950 complete with bayonets in September 1745. At the time of writing a single example has been identified and is now in the Royal Armouries Museum at Leeds. The workmanship, as with other Barber / Ordnance weapons of this period, is markedly inferior to that of regular contract arms.

The Duke of Cumberland's Dragoon Carbine of 1746

George II's second (and favourite) son, William, Duke of Cumberland, was allowed to raise a regiment of Dragoons for service on the Continent in 1746. For reasons not revealed in the records of the time, 'H. R. H.' as he was generally referred to, did not arm his unit with the new Short Land Musket with Wood Rammer for Dragoons which had gone into production in 1744, but turned instead to the private sector for his regiment's arms. The Duke may have wished to stress the 'light' element of his unit and preferred a carbine-bore weapon to the larger and heavier musket recently introduced for the regular dragoons and dragoon guards, who had always been equipped with musket-bore longarms. Whether or not influenced by the Master General of the Ordnance, or by Thomas Hatcher, the new Master Furbisher, who visited the Duke at Kensington Palace *'with Pattern Arms'* in September, and *'Pattern Pistols for dᵒ.'* in October 1746, the Duke turned to James Barber who supplied 390 carbines in carbine-bore with socket bayonets in

November-December 1746, with a second delivery of 30 in December 1747 and a final 68 in December 1748, for a total of 488. The majority were lost through shipwreck and battlefield casualties, and no example is known at the time of writing.

Pattern 1748 Extraordinary Land Service Muskets

The term 'extraordinary' as used by the Ordnance seems to relate most frequently to length and sometimes to weight or finish (as with flat-faced carbine and pistol locks from 1756). It was a far more common term in Ordnance usage during the last decade of the Seventeenth and first decade of the Eighteenth Century, and this is one of its last appearances amongst Ordnance small arms. Certainly every component of these unknown muskets, with the exception of the brass furniture, cost more than the same piece of a conventional Long Land Pattern musket. Acting on a warrant dated 6 September 1740, Edward Jordan provided 100 Extra barrels at 12/- each later that same month, but it was another seven years before any further steps were taken to complete these arms. At the time a standard Long Land musket barrel cost 7/6. John Smith provided all 100 double-bridle locks for these muskets at 10/- each in March 1747 to an order dated December 1746; the price of a current musket lock was 6/-. Richard Waller rough-stocked all 100 at 8/- each in June 1747, when he was paid 2/4 for rough-stocking a Long Land musket, suggesting considerable extra work and/or wood being involved. In September 1747, Thomas Hollier provided 100 sets of Extraordinary brass musket furniture at the same cost as a set of conventional furniture. Finally, after the war had come to an end, James Barber and Charles Pickfatt each set up 50 at 9/- each in September 1748, when they were being paid 7/8 for a conventional Land Service musket. These muskets therefore cost the Ordnance 15/6 more than a conventional infantry musket, a considerable increase for the time.

Given that wall pieces were an article in the standard production line, and that 'Long Fowlers' had been produced as recently as 1738, the naming, the appearance and the purpose of these Extraordinary muskets remain a distinct puzzle. Given the timing of the original component order, they were most likely intended for some activity connected with the colonial operations of the War of Jenkins' Ear against Spain in the Caribbean, some operation which was either completed or abandoned before further progress could be made with the muskets. Why they should have been completed seven years later, after the end of the war, is another unanswered question.

Fusils for Royal Artillery Officers, Pattern 1750

Since equipping part of the artillery crews with carbines in the course of 1748, the idea developed that officers too should carry something more effective than a sword. As with the cavalry and infantry, so there was within the Royal Artillery an on-going argument over whether firearms were appropriate to the equipment of officers who were supposed to spend their time in battle leading, encouraging and controlling their troops, not shooting at the enemy. However, between 1750 and 1772 the firearms people won, and artillery officers were issued with fusils made by the workforce in the Small Gun Office in the Tower of London.

In June 1750, Thomas Hatcher, the new Master Furbisher, received the Surveyor General's orders to prepare carbines sufficient to equip the officers of five companies of artillery, and, accordingly, ordered parts for 38 carbines. These were reported completed and turned in to Store with their bayonets and buff slings in August 1750. A further 7 were ordered made at the end of 1751, and another 3 in April 1753. A further 16 were set up by John Hirst in 1756, but there was no further production until 50 were set up in 1762. Further production occurred in 1765, 1767 and a final run of 119 in 1770. Two years later, officers were required to give up their fusils and receive a new pattern of sword. Total production amounted to 321 fusils.

At the time of writing no example of this carbine has been identified. They have conventional carbine-bore barrels of 36" length to take a socket bayonet, but are fitted with a false, or break-off, breech, the tang of which and the brass furniture are engraved. Complementary to the break-off breeches, the barrels are secured to the stocks with slides or 'bolts' as the Ordnance termed them, and the wooden rammers are fitted with a worm.

The Indian Fusils of 1753

After a gap of thirty-seven years, Thomas Hartwell, Clerk of the Ordnance and Assistant Master Furbisher, gun designer and modeller, was given charge of an order for 400 Indian fusils to be presented by the British Government to the Six Nations (Iroquois) in the Province of New York. This was the first Ordnance-produced group of Indian guns since 1716, and was to be the last until the 1813 series. This 1753 order is of particular interest and importance because it contains the first use noted of the term 'Northwest' to denote a type of barrel for Indian guns, and because it is the first recorded instance in which two distinct grades of guns, later to be known as 'Chief's' and 'Common' guns, were made by the Ordnance for British Government use. Whether the Ordnance originated this division in quality or took the idea from the pre-existing commercial Indian trade is not known. The Surveyor General of the Ordnance ordered Hartwell to prepare pattern guns of two sorts suitable for Indians, and in August 1753 these were produced and approved. They were to be made by the Piecemen in the Small Gun Office, at a total cost of £220.

Thus we find Hartwell ordering the following components:—

20 Best Northwest 4 foot 0 in Barrels 6/6 each	380 setts of flatt d°. - 10d each
380 four feet ditto made to pattern - 4/- each	For engraving 20 best guns - 1/6 each
20 walnuttree Stocks & Stocking - 5/- each	For engraving 380 common guns 8d each
380 mottled Beech stocks &c. - 2/- each	For 410 four feet ash rammers
20 setts birding Brasswork - 1/6 each	177.0.6. total

Note that the barrels, for the first time on Ordnance-made Indian guns, conform to the popular commercial length of 48-inches rather than the Ordnance length of 46-inches used previously. The absence of any mention of locks suggests that these were already available to the setters up, probably Land Service carbine locks from Store.

After 1753, and with the exception of the 1813 Indian arms (see below), Indian guns for use by the British Government were obtained directly from the commercial trade, and until at least the 1790s almost exclusively from the Wilson firm in the Minories, London.

Fusils for Marine Officers by Freeman and Barber, 1755 – 1763

When the Marines were again raised beginning in April 1755, arms of all sorts were an immediate priority, since every private and corporal was to have a musket and bayonet. An earlier manufacture of arms for marines in 1748 had been completed too late in the war to be issued, but their numbers were inadequate to the numbers now required, and the evidence suggests that they were issued for sea service prior to the new-raising of the marines. But the idea of fusils for marine officers was a new one, doubtless encouraged by the design and issue of fusils to artillery officers, and their use by infantry officers in the coming war.

Total production by James Freeman was 48 (there is some doubt whether he actually delivered them) in 1755 only, and by James Barber, 951 from 1755 to 1763. None have been identified at the time of writing. However, like most of the earlier complete arms supplied by contractors they are very likely to be signed by their respective makers and to bear the King's Proof Marks on the breech of the barrel and an Ordnance Storekeeper's stamp on the butt, and to represent a compromise in the quality of both components and finish.

Blunderbuss-carbines for Gen. Burgoyne's 23rd Light Dragoons, 1781

Having returned to England after his defeat and capture at Saratoga in 1777, Gen. John Burgoyne became involved in designing arms for his newly raised 23rd Light Dragoons who were ordered to India. Burgoyne designed a small carbine with a barrel having an elliptically flared muzzle, and a pistol which was apparently also well wide of the current Ordnance Light Dragoon pattern. The Master Furbisher *'reported that it was much inferior'* to the present pattern and no more is heard of it. The little carbine was approved and 100 were ordered made in October, of which 60 were ordered to be issued by a Royal Warrant of 3 November 1781. In fact, no more than 60 were produced unless the balance of 40 were made by the Small Gun Office workforce, of which there is no mention or other evidence.

Samuel Galton made the 16⅛ inch carbine-bore browned and flared barrels ($1^{7}/_{16}$" x 1" at the muzzle) for 7/- each. James Hirst and Alexander Davidson shared the setting up of 30 each at 9/- for which they billed the Ordnance in January 1782. The lock is the Pattern 1777 Extra Flat Carbine lock, and the furniture includes the notched nosecap and swell & groove rammer, buttplate and trigger guard of the Eliott Carbine, with a variant but similar sidepiece, and a short sling bar on the left side fastened in the same manner as the Royal Foresters' Carbine. The barrel is held by two keys. At the time of writing only one example is known, in the Royal Armouries Museum, Leeds.

Pistols for the Duke de Bouillon's organization on the Channel Islands, 1796

A total of two hundred pairs of these medium-sized brass-barrelled and brass-mounted pistols were produced in equal quantities as complete weapons by London gunmakers Durs Egg and Henry Nock, and delivered in May 1796 at 48/- a pair. They were made for the use of Philip d'Auvergne, Duke de Bouillon, a Royal Navy Captain who operated a communications / spy / privateering network on the islands of Guernsey and Jersey off the French coast. While Egg's production has only his name engraved across the tail of the lockplate, Nock's have both his name and TOWER stamped in the same position. The seven-inch brass barrels bear the King's Proof Marks.

Pattern 1800 Duke of York's Hussar Carbine

The only British flintlock regulation carbine to be made with a 21-inch barrel was first produced by Henry Nock, who delivered 504 in August 1800. The steel rammers were made by him from cut-down India Pattern musket rammers. Nock supplied all of the components except the barrels, at a cost of 20/4 each to the Ordnance.

The Ordnance prepared to make additional carbines of this pattern, since 761 sets of brass furniture for it were delivered between April and September 1805, and James Bailey delivered 560 sets of smallwork for them, also in 1805. These were utilized in the Royal Manufactory in the

Tower to produce 259 Duke of York's Hussar carbines that year, and it seems likely that the balance were completed during 1806, although there are no production figures for the Manufactory for that year. These carbines were issued to the cavalry of the King's German Legion, and at the time of writing no example has been identified.

The Pattern 1801 Life Guards' Rifle

Only forty-eight of these rifles were delivered in September 1801, all by Ezekiel Baker. Of the total, forty were delivered with the locks being supplied by the Ordnance, for 89/- each, and the remaining eight complete with locks from Baker for 97/-, the same price as a complete Infantry Rifle. These were the first rifled arms designed specifically for a British mounted regiment and were clearly intended as a trial. Speculation suggests that these rifles may have a sling bar mounted on the left side of the stock opposite the lock, and that they were probably not fitted with a bayonet. It is probable that the forty locks supplied by the Ordnance will have conventional markings, while the eight from Baker may have his own marks. All should have the King's Proof Marks and Storekeeper's stamp in the butt.

The Pattern 1810 Musket-bore Infantry Rifle

All 100 of these ·70 calibre rifles were rough-stocked and set up, complete with box implements, by R. & R. Sutherland, and delivered into Ordnance Stores in three parcels, April to June 1810. The one identified example (in the Royal Armouries Museum, Leeds) exhibits a combination of old-fashioned and current design features. Amongst the old-fashioned are the large butt-box with a stepped border to the hinge-piece unique to this pattern. There is no hint in the surviving records why the Ordnance was willing to produce this special design weapon at this date. Production of 100 suggests a company-size armament, with perhaps a few spares for replacements. Why had the musket-bore been resurrected after its clear unpopularity and rejection only nine years earlier?

The Pattern 1812 Life Guards Rifled Carbines

No example of this carbine has been identified at the time of writing, and the vague and imprecise references to its design favour a weapon which will closely resemble a Pattern 1798 Heavy Dragoon Carbine and bayonet, of musket-bore but with a rifled (almost certainly Baker seven-grooved) bore, a twist-iron barrel and presumably a backsight, most probably a plain notched block like the current Cavalry Rifle. A total of 102 was produced by only eight Birmingham contractors, who delivered them into Store between May and September 1812. It is curious to note, however, that in December 1814, an enquiry into the cost of these carbines was ordered – were they by then also being produced in the Royal Manufactory in the Tower?

The Indian Guns of 1813

The largest group of Ordnance-sponsored Indian guns was produced by eighteen Birmingham contractors (noted in their entries in Part I) who delivered Chief's and Common guns, Rifles and Pistols intended for gifts to the Canadian and other British-allied Indians in the Gulf Coast area of the southern United States during the War of 1812. Each of these types closely followed the current designs in use by the commercial trade in Indian goods, and most were made by the same makers who served the civilian trade. A total of 26,801 arms were produced between 1813 and 1816, most of them arriving too late to serve their intended purpose. In contrast to the civilian-market arms, these will bear the King's Proof on their barrels and often be branded on the stocks with a Broad Arrow over BO or sometimes I D [Indian Department]. For a detailed discussion of this series see my article '*Those Board of Ordnance Indian Guns-Again!*' in The Museum of The Fur Trade Quarterly, vol. 21, No.1 (Spring 1985) and No. 3 (Fall 1985).

The Pattern 1814 Russian Pistols

One of the most intriguing productions of the later flintlock period, only one example of what may have been more than five thousand produced, has been identified, in the collections of the Royal Armouries Museum, Leeds. Contractor production was restricted to seventeen Birmingham makers and began in January 1814, but references in the records strongly suggest that production was also subsequently undertaken by the Royal Manufactory of Small Arms in the Tower of London, who may have produced the great majority of these strange pistols. By December 1813, there were reportedly 3,400 in process of manufacture, and in mid-1814, six months after initial deliveries, 2,268 brass barrel bands and '*Caps with studs*' as well as front sights were delivered. They were last being delivered in March 1816.

The above group of specialist arms, manufactured in almost all cases by the contractors who supplied regulation pattern small arms to the Board of Ordnance, demonstrates the diversity of arms which the Ordnance was willing to see created under their supervision, either to meet a requirement not covered by the regulation patterns, or to prevent the imposition of higher prices and deficient designs and/or workmanship on some unit commander who managed to create a sufficiently strong case to be specially catered for, or who wielded sufficient influence to achieve the same end in the face of ever-increasing consolidation and centralization of Britain's military administration.

PART 1

Comprehensive Alphabetical Listing

NOTES: The years during which each contractor worked for the Board are shown as: ID Mar. 1715 and FD Jan. 1721. These represent 'initial bill date in Bill Book (or Treasurer's Ledger warrant date)' and 'final date of appearance in either of the above sources.' Where the ampersand (&) is used- as in Rough Stocked & Set Up, it indicates that the two operations so connected were carried out successively on the same group of guns by that contractor, and where 'and' is used it indicates that each operation was carried out by itself, and not necessarily on the same group of guns. In a few instances, particularly from 1805 to 1816, the nature of the documentary entry may not include a month because detailed accounting for component parts was discontinued with the introduction of open contracts, but every effort has been made to locate as accurate and precise a date as possible. Continuing research may supply some of these omissions. In general the end of the month mentioned may be assumed.

The use of a semi-colon between two dates indicates deliveries only at those dates. Two dates with a dash between them indicates a series of deliveries during the time between the dates.

Contractors with Initial Dates of October 1688 may be assumed to have worked during the reign of James II. Some of the arms furnished under this contract represent the last of the regularly supplied matchlock muskets procured by the Ordnance.

A

ADAM, Alexander. London. Aug. 1798 only.
MERCHANT who supplied complete arms purchased in the open market. Also ROUGH STOCKED & SET UP one batch of India Pattern muskets.

ADAMS, John. London. ID July 1825 ; FD July 1826.
BARRELS and ROUGH STOCKED & SET UP, Pattern 1823 Infantry Rifles.

ALDRIDGE, George. London. ID July 1705 – FD June 1716.
COMPLETE ARMS to June 1712, BARRELS for Irish service muskets in Dec. 1713. STOCKER & SETTER UP from Sept. 1714. Supplied Government-gift New York Indian fusils in 1710 and 1716, and cleaned and repaired some of the latter after their loss by shipwreck and recovery.

ALLPORT, William. Birmingham. ID Apr. 1816 – May 1819; FD Nov. 1823.
LOCKS, BAYONETS, COMPLETE ARMS, including 1813 Indian arms, chiefly New Land Pattern Infantry and Light Infantry muskets. Infantry Rifle locks only in 1823.

AMPHLETT, Richard. Birmingham. ID Aug. 1795 – FD 1796.
BAYONETS, carbine.

ARCHER, Thomas. Birmingham. ID Apr. 1794 – FD June 1805.
BARRELS and a small number of COMPLETE ARMS.

ARCHER & SON, Thomas. Birmingham. ID June 1805 – Feb. 1818.
BARRELS and COMPLETE ARMS.

ARKILL (ARKELL), Peter. London. ID Mar. 1696 – FD Dec. 1706.
COMPLETE ARMS.

ARNOLD, Francis. Birmingham. ID Mar. 1806 – FD Nov. 1811.
BARREL RIFLER, sometimes specifies 30-inch barrels.

ASHMORE, R. Birmingham. Feb. 1838 only.
LOCKS, Pattern 1823 Infantry Rifle.

ASHTON, J. Birmingham. Feb. 1838 only.
LOCKS, Pattern 1823 Infantry Rifle.

ASHTON, Thomas. London. ID Oct. 1804 – Mar. 1818; 1835; Aug. 1836 – FD Jan. 1839.
BARRELS, ROUGH STOCKED & SET UP, and COMPLETE ARMS. 16" Cavalry Carbines for the Queen of Spain in 1835, Pattern 1823 Infantry Rifles for the Shah of Persia in 1836, pistols 1835 AND 1837, carbines 1835-7, and a final delivery of infantry rifles in 1838-9.

ASHTON, Thomas & Charles. London. ID Feb. 1824; 1825; FD Apr.1827.
ROUGH STOCKED & SET UP, Pattern 1823 Infantry Rifles.

ASPINALL, James. Birmingham. ID Aug. 1812; FD Sept. 1812.
BARRELS.

ASTON, Joseph. Birmingham. July 1838 only.
IMPLEMENTS, Infantry Rifle.

ATHERTON, John. Birmingham. Dec. 1795 only.
BAYONETS, musket.

AUSTIN, Jacob. Birmingham. ID Mar. 1699 – FD June 1709.
COMPLETE ARMS and deliveries of LOCKS (1707 – 09). A contract dated 1693 mentions Jacob Austin, but the above ID is his first bill in the Ordnance records after 1689.

AUSTIN, Mary. London. ID July 1710 – FD June 1711.
COMPLETE ARMS. Widow of Thomas.

AUSTIN, Thomas. London. ID Oct. 1688 – FD Mar. 1709.
COMPLETE ARMS. BARRELS and LOCKS (Nov. 1699 only). A major supplier. Supplied Government-gift New York Indian fusils in Aug. 1699, June 1700 and Oct. 1702.

B

BAILEY, James. London. ID June 1803 – FD 1823.
SMALLWORK, LOCKS (1808). Virtual monopoly supplier of small-work to the Board.

BAKER, Ezekiel. London. ID Apr. 1794 – FD Nov. 1825.
Until Mar. 1798 supplied only cut, new looped and sighted BARRELS. Thereafter supplied ROUGH STOCKED & SET UP and COMPLETE ARMS, and acted as a barrel rifler. Submitted the rifle barrel adopted in 1800 for the new Infantry rifle and subsequently (1803) the Cavalry rifle.

BAKER, Ezekiel John. London. ID Aug. 1836 – FD Mar. 1839.
ROUGH STOCKED & SET UP. Pistols and carbines, and Pattern 1823 Infantry Rifles for the Shah of Persia in 1836; carbines in 1836 – 7; infantry rifles in 1838 – 9. Son of Ezekiel, and succeeded the partnership of himself and his mother Elizabeth.

BAKER & SON, Elizabeth. London. ID Mar. 1826 – FD 1835.
ROUGH STOCKED & SET UP, COMPLETE ARMS and IMPLEMENTS, Infantry and Cavalry Rifles, and 16" Cavalry Carbines for the Queen of Spain in 1835. Successor to Ezekiel.

BANBURY, William. London. ID June 1715 – FD Mar. 1716.
ENGRAVER of locks.

BANKS (BANCKS), Henry. London. ID Oct. 1688 – FD June 1700.
COMPLETE ARMS. BARRELS and LOCKS (Nov. 1699 only).
Supplied Government-gift New York Indian fusils in Aug. 1699 and June 1700.

BANISTER, Robert. London. ID June 1702 – FD Nov. 1708.
COMPLETE ARMS.

BANISTER, Thomas. London. ID Oct. 1688- FD Jan. 1705.
COMPLETE ARMS; one delivery of BARRELS (Nov. 1699) and LOCKS (1703).

BARBER, James. London. ID Mar. 1741 – FD June 1749.
SETTER UP and COMPLETE ARMS. Supplied complete one delivery of pistols for Churchill's Dragoons (1745) and all of the carbines for Lord Loudoun's Regiment of light infantry (1745) and the Duke of Cumberland's Dragoons 1746, 1748. Successor to his father Lewis.

BARBER, Lewis. London. ID Sept. 1722–3; 1725; 1727; Apr. 1730 – FD July 1741.
SETTER UP and COMPLETE ARMS. One of the two setters up to the Board during his operational span, along with Charles Pickfatt. Supplied a series of nineteen pattern muskets to the Master General of the Ordnance in 1722, none of which were subsequently adopted. Also supplied several complete orders of 'regimental' arms: Dec. 1723 muskets and pistols for the Duke of Bolton's [Royal Horse Guards], Col. Campbell's [2nd or Royal Scots Greys], Lord Carpenter's 3rd Dragoons, and for Brigadier Gore's 1st, Royal Dragoons and Col. Kerr's regiment of foot; in Dec. 1727, muskets and bayonets for Maj. Gen. Grove's 10th Foot, Brig. Kerr's 13th Foot, and muskets and bayonets and pairs of pistols for Col. James Campbell's 2nd Dragoons and Carpenter's 3rd Dragoons; in 1740, muskets for Lord Crawford's regiment (the original Highland Regiment, subsequently the 42nd). Succeeded by his son James.

BARKER, Matthias. Birmingham. ID Sept 1775 – FD Nov. 1782.
BARRELS supplied in partnership with John Whately. One of four contractors who supplied COMPLETE Pattern 1776 muzzle-loading, and breech-loading Ferguson rifles to the Board during 1776.

BARKER & HARRIS. Birmingham. ID June 1771 – FD May 1775.
BARRELS. Partnership of Matthias Barker and widow Jane Harris.

BARNES, John. London. ID June 1719 – FD Mar. 1720.
SETTER UP of Pattern 1718 muskets.

BARNETT, John Edward. London. ID June 1832; 1835 – FD May 1839.
    ROUGH STOCKED & SET UP, including pattern 'Manton' carbines in 1832. 16" Cavalry Carbines for the Queen of Spain in 1835-6; pistols in 1836–7; carbines 1835–7, infantry rifles in 1838–9.

BARNETT, Thomas. London. ID Aug. 1793 – FD Mar. 1818.
    ROUGH STOCKER & SETTER UP, and COMPLETE ARMS. Also supplied non regulation pattern arms purchased in the open market in the 1790s.

BARNETT & SON, Thomas. London. ID Mar. 1818 – 1819; Nov. 1820 – 1821; Mar. 1824 – FD Jan. 1827.
    ROUGH STOCKED & SET UP and COMPLETE ARMS including Infantry Rifles in 1820, 1824, 1825 and 1827, pistols in 1821, African muskets without bayonets @ 21/- and 'Tower Guns' @ 16/- in 1824. These trade arms should have Ordnance markings.

BARRAS, Ralph. London. ID Dec 1718 – FD June 1720.
    SETTER UP of the Pattern 1718: three deliveries.

BATE, Edward. London. ID Dec. 1779 – FD Mar. 1783.
    COMPLETE ARMS (Sea Service) and SET UP some Short Land muskets, and carbines.

BATE, Thomas. Birmingham. ID Dec. 1807 – FD Dec. 1816.
    BAYONETS, particularly rifle sword-bayonets.

BATTY, Henry. London. ID June 1709 – FD Mar. 1710.
    COMPLETE ARMS.

BAYLISS, Joseph. Birmingham. ID Mar. 1798 – FD Feb. 1811.
    BARRELS.

BAYLISS, Lewis. Birmingham. Feb. 1824 only.
    LOCKS, Pattern 1823 Infantry Rifle.

BAYLISS & SON, Joseph. Birmingham. ID Mar. 1811 – FD 1824.
    BARRELS.

BECKWITH, William A. London. July 1838 only.
    Boring, rifling, sighting, smoothing, browning and fitting hand-bayonets to Infantry Rifle barrels.

BENNETT, John. London. ID Feb. 1691 – FD Mar. 1711.
    COMPLETE ARMS. BARRELS and LOCKS (Nov. 1699 only).

BIGGLESTONE, Thomas. London. ID Apr. 1702 – FD Aug. 1711.
    BAYONETS.

BILLS, Richard. Wednesbury. ID Sept. 1823 – Aug. 1826; FD Feb.1828.
    LOCKS, Pattern 1823 Infantry Rifle, and Pattern 1798 Heavy Dragoon Pistol in 1828 for RSAF Enfield.

BILLS, Samuel. Wednesbury. ID Mar. 1804 – Feb. 1818; Sept. 1823 – FD Sept. 1826.
    LOCKS. Pattern 1823 Infantry Rifle only in 1820s.

BIRD, William. Birmingham. ID Apr. 1790 – FD Aug. 1795.
    LOCKS and BARRELS.

BIRKELL, William. London. ID July 1742 – FD June 1749.
SETTER UP. The most important, in terms of production, of eight firms, brought in by the Board during 1742, to increase the output of small arms.

BIRKLEY, Thomas. London. Oct. 1794 only.
SETTER UP of one order of Short Land muskets.

BISSELL, Isaac. Birmingham. ID Aug. 1779 – FD Aug. 1783.
BARRELS.

BLAIR, David. Birmingham. ID Nov. 1793 – Mar. 1804; Feb.1809 – FD May 1814.
LOCKS, BARRELS, ROUGH STOCKED & SET UP and COMPLETE ARMS, including 1813 Indian arms. His widow, Jane Hannah, took over the business in June 1814.

BLAIR, Jane Hannah. ID June 1814 – Sept. 1819; Nov. FD Dec. 1823.
ROUGH STOCKED & SET UP, COMPLETE ARMS, LOCKS and BARRELS. Also 1813 Indian arms. Widow of David Blair, took over his contracts. Pattern 1823 Infantry Rifle barrels and locks, 1823.

BLAIR & SUTHERLAND. Birmingham. ID Mar. 1804 – FD Jan. 1809.
BARRELS, LOCKS, ROUGH STOCKED & SET UP and COMPLETE ARMS and BAYONETS. Partnership of David Blair and the Sutherland brothers, Ramsay and Richard. Partnership dissolved and each continued to deliver separately.

BLAKEMORE, J.P. Darlaston. ID Oct. 1823 – Dec. 1826; FD Mar. 1828.
LOCKS, Pattern 1823 Infantry Rifle; Pattern 1798 Heavy Dragoon Carbine and Pistol for RSAF Enfield in 1828.

BLAKEMORE, Mary. Darlaston. ID Sept. 1823 – FD Dec. 1826.
LOCKS, Pattern 1823 Infantry Rifle.

BLAKEMORE, Thomas. Darlaston. ID June 1775 – Oct. 1781; Dec. 1793 – FD Feb. 1818.
LOCKS. Replaced the late George Haskins as lock contractor in Oct. 1770. Probably two generations of the same name.

BLAKEMORE & ROBBINS. Birmingham. Mar. 1807 only.
LOCKS. Successors to Blakemore & Son.

BLAKEMORE & SON, Thomas. Darlaston. Apr. 1807 only.
LOCKS.

BLAKEMORE, Thomas W. Darlaston. Apr. 1815 only.
LOCKS. Probably the son of Blakemore & Son.

BLANKLEY (BLANCKLEY), John. London. ID Oct. 1688 – FD Nov. 1699.
COMPLETE ARMS. BARRELS and LOCKS (Nov. 1699 only).

BLANCKLEY, Samuel. London. ID Sept. 1703 – FD Oct. 1706.
COMPLETE ARMS. Three deliveries of Pattern 1703 muskets.

BLEAMIRE (BLAYMER, BLAMYER), Winifred. London. ID Oct. 1688 – FD Apr. 1702.
COMPLETE ARMS. BARRELS and LOCKS (Nov. 1699 only).

BOND, William James. London. Aug. 1836 only.
ROUGH STOCKED & SET UP, Shah of Persia Infantry Rifles.

BOND, William Thomas. London. ID 1835 – FD 1837.
ROUGH STOCKED & SET UP, 16" Cavalry Carbines for the Queen of Spain in 1835–6; carbines and pistols in 1836–7, and Pattern 1823 Infantry Rifles for the Shah of Persia in 1836. Supplied a small order of COMPLETE single-barrelled fowling pieces and pairs of pistols for Portendik (West Africa) native chiefs in July 1837.

BONUS & CO. (BONUS & HOLBROOK). Birmingham. Jan. – June 1796.
COMPLETE Sea Service muskets and one delivery each of locks and barrels.

BOULTON, Peter. London. ID Oct. 1688 – FD May 1714.
COMPLETE ARMS to Sept. 1711. BARRELS and LOCKS (Nov. 1699), and BARRELS and LOCKS for Irish service muskets May 1714.

BOURNE, William. London. Mar. 1700 only.
COMPLETE ARMS. One delivery.

BOWERS, George. London. ID May 1704; FD Mar. 1705.
COMPLETE ARMS. Two deliveries.

BRANDER, Martin. London. ID Dec. 1793 – FD Nov. 1809.
ROUGH STOCKER & SETTER UP, and COMPLETE ARMS. The most important London contractor for supplying India Pattern Muskets. Became Brander & Potts.

BRANDER & POTTS. London. ID Dec. 1809 – May 1819; 1820 – 1821; Feb.1824 – FD Mar. 1827. ROUGH STOCKED & SET UP and COMPLETE ARMS. Partnership of Martin Brander and Thomas Potts. Infantry rifles in 1820 and 1824-7, pistols in 1821.

BRAZIER, John. London. ID May 1756 – FD June 1757.
SETTER UP.

BRAZIER, William. London. ID Feb. 1714 – FD June 1716.
SETTER UP.

BRIDGER, George. London. ID Dec. 1703 – FD Dec. 1707.
COMPLETE ARMS.

BROOKE, John senior. London. ID Dec. 1701 – FD Mar. 1716.
COMPLETE ARMS to Nov. 1710. STOCKER & SETTER UP Mar. 1715 - Mar. 1716.

BROOKE, John junior. London. ID Mar. 1715 – FD June 1718.
STOCKER & SETTER UP.

BROOKE, Mary. London. ID Oct. 1688 – FD Feb. 1692.
COMPLETE ARMS.

BROWN, Henry. Birmingham. July 1805 only.
BARRELS, one delivery of India Pattern musket barrels.

BRUSH, John. London. ID June 1702 – FD Aug. 1715.
COMPLETE ARMS to Mar. 1713, including Government-gift New York Indian fusils in Feb. 1710. Also LOCKS in 1708-09.

BUCKMASTER, Joseph. London. ID July 1756 – FD July 1757.
SETTER UP.

BUCKMASTER, Robert. London. ID Aug. 1719 – FD Dec. 1720.
SETTER UP of Pattern 1718 muskets.

BUCKMASTER, William. London. Apr. 1720 only.
SETTER UP of Pattern 1718 muskets: one delivery.

BULLEIS (BULLESS), Thomas. London. ID Dec. 1810 – Nov.– Dec. 1823
LOCKS and LOCK FORGINGS (Feb. 1811); locks and barrels for Pattern 1823 Infantry Rifles in 1823.

BUMFORD, John. London. ID July 1756 – July 1757; FD Mar. 1771.
SETTER UP. One delivery of wall pieces complete with moulds in 1771.

BUNDAY, John. London. ID Oct. 1688 – FD June 1718.
COMPLETE ARMS to May 1711 including Government-gift New York Indian fusils in Feb. 1710, and LOCKS for Irish service muskets in Nov. 1713. STOCKER & SETTER UP from Mar. 1715.

BURGIN, Mary. London. ID Dec. 1718 – June 1719; May – FD Dec. 1727.
BRASS GUN FURNITURE. Widow of William.

BURGIN, William. London. ID Jan. 1714 – FD Sept. 1716.
BRASS GUN FURNITURE.

BYE, Richard. London. ID May 1703 – FD Mar. 1710.
COMPLETE ARMS including Government-gift New York Indian fusils in Feb. 1710.

BYE, Sarah. London. ID July 1710 – FD Sept. 1713.
COMPLETE ARMS to Dec. 1712, and BARRELS for Irish service muskets in Sept. 1713.

C

CADDY, Edward. London. ID Oct. 1688- FD Sept. 1707.
COMPLETE ARMS. BARRELS and LOCKS (Nov. 1699 only).

CAM, James. Birmingham. ID July 1795- FD 1796.
BAYONETS.

CARLYON, Nicholas. London. ID Dec. 1703- FD Mar. 1709.
COMPLETE ARMS.

CARTER, Benjamin. London. ID June 1718- FD Mar. 1721.
STOCKER & SETTER UP of Pattern 1718 arms.

CASLON, William. London. ID July 1716 – FD May 1719.
ENGRAVER of locks. At the time of writing the only Ordnance lock engraver known to have signed his work (internally).

CASTLE, Edmund. London. ID Dec. 1701 – FD July 1707.
COMPLETE ARMS.

CHAMBERS, Samuel. Birmingham. ID Aug. 1812 – FD Feb. 1818.
BAYONETS.

CHAMPANTE (CHAMPANTIE), Henry. London. ID June 1702 – FD Dec. 1703.
COMPLETE ARMS, LOCKS. Two deliveries.

CHASE, Samuel. London. ID Dec. 1703 – FD Aug. 1711.
BAYONETS.

CLARKE, William. London. ID Oct. 1739 – FD Jan. 1749.
  BARRELS and LOCKS, SMALLWORK, BULLET MOULDS. Also delivered 500 sockets for throwing handgrenadoes, Mar. 1742. After 1742 almost entirely concerned with repairing and re-working barrels.

CLARKSON, Joseph. London. ID Dec. 1723 – FD Mar. 1728.
  COMPLETE ARMS. Regimental colonel's muskets and bayonets, and pistols delivered in Dec. 1723 to: the Earl of Stair's [6th] and Col. Churchill's [10th] Dragoons.

CLIFFORD, Peter. London. ID Mar. 1708 – FD Dec. 1709.
  LOCKS, musket.

CLIVE, John. Birmingham. ID Sept. 1825; FD Sept. 1838.
  BARRELS, Pattern 1823 Infantry Rifle.

CLIVE, John junior. Birmingham. Feb. 1824 only.
  LOCKS, Pattern 1823 Infantry Rifle.

CLIVE, Thomas. Birmingham. Dec. 1823 only.
  BARRELS, Pattern 1823 Infantry Rifle.

CLIVE & SON. Birmingham. ID May 1816 – FD Feb. 1818.
  BARRELS.

CLIVE & TURTON. Birmingham. ID July 1811 – Dec. 1816; Aug. 1823 – FD Dec. 1825.
  BARRELS, Infantry Rifle only after 1816. Partnership of John Clive and Joseph Turton.

COLE, Elias. Birmingham. ID Feb. 1710 – FD May 1734.
  BARRELS (1710; 1714 – 1723 including 400 for Indian fusils 1716; 1726-34).
  LOCKS (1716 – 1720 including 400 for Indian fusils 1716; 1728 – 1730), and BAYONETS (1718 – 1722; 1727). Major supplier.

COLE, Thomas. London. ID Apr. 1691 – FD Nov. 1694.
  COMPLETE ARMS.

COLLETT, Joseph. London. ID Dec. 1718 – FD Mar. 1721.
  STOCKER & SETTER UP of Pattern 1718 muskets.

COOKES, Edward. Birmingham. ID Nov. 1707 – Mar. 1712; Dec. 1718 – June 1719; Sept. 1727 – FD Jan. 1741. LOCKS (1707-19; 1729), BARRELS (1711; 1727-41). ENGRAVER (1707 – 1715). Major supplier.

COOPER & CRAVEN. Birmingham. ID Nov. 1803 – FD 1819.
  BAYONETS and SWORDS.

CORBETT, J. Birmingham. June 1838 only.
  LOCKS, Pattern 1823 Infantry Rifle.

CRAVEN, Thomas. Birmingham. ID Sept. 1802 – FD July 1803.
  BAYONETS, SWORDS, SPEARS.

CRIPS (CRIPPS), Henry. London. ID Oct. 1688 – FD June 1706.
  COMPLETE ARMS including Government-gift New York Indian fusils in 1699, 1700 and 1702. BARRELS and LOCKS (Nov. 1699), and alterations, modifications such as ribbing and banding carbines and dragoon muskets, fitting socket bayonets and sighting barrels, and repairing work. Succeeded George Fisher as Furbisher in the March Quarter of 1695, and was succeeded at his death by Richard Wooldridge in the Spring of 1710.

CROOKE, Robert. London. ID Feb. 1691 – FD Nov. 1693.
   COMPLETE ARMS. Three deliveries.

CROSBEE, William. London. ID Mar. 1812- FD Sept. 1814.
   BRASS GUN FURNITURE.

D

DARKE, Elizabeth. London. ID June 1702 – FD Dec. 1709.
   COMPLETE ARMS.

DARKE, Harward. London. ID Oct. 1688 – FD Mar. 1696.
   COMPLETE ARMS. One Sea Service delivery in 1701 to Thomas Darke in Harward's name.

DARKE, Thomas. London. ID June 1709 – FD Dec. 1712.
   COMPLETE ARMS.

DAVIDSON, Alexander. London. ID Dec. 1781 – FD Mar. 1786.
   SETTER UP and BARREL FILER.

DAVIS (DAVIES), Thompson. London. ID Jan. 1777- FD Apr. 1780.
   SETTER UP, Land Pattern muskets.

DAVIS, Robert. London. ID May 1719 – FD Mar. 1720.
   ROUGH STOCKER of Pattern 1718 muskets.

DAVIS, William. London. ID June 1702 – Dec. 1706; Nov. 1718 – FD Oct. 1719.
   COMPLETE ARMS 1702 – 1706. STOCKER & SETTER UP of Pattern 1718 arms.

DAVIES & CO., William. Birmingham. ID Oct. 1803 – FD Feb. 1808.
   BARRELS.

DAWES, John. Birmingham. ID June 1816 – FD May 1819.
   BARRELS, BAYONETS, ROUGH STOCKED & SET UP, New Land Pattern muskets. Took over from Samuel & John.

DAWES, Samuel & John. Birmingham. ID June 1813 –July 1816; FD Dec.1823.
   BARRELS, BAYONETS and ROUGH STOCKED & SET UP.
   Successors to William & Samuel. Pattern 1823 Infantry Rifle barrels in 1823.

DAWES, William. Birmingham. ID Feb. 1779 – Nov. 1782; Nov. 1794 – FD 1798.
   BAYONETS.

DAWES, William & Samuel. Birmingham. ID June 1798 – FD May 1813.
   BARRELS, BAYONETS, ROUGH STOCKED & SET UP and COMPLETE ARMS including 1813 Indian arms. Sometimes signed as Dawes & Son. William dead May 1813.

DEAKIN, Francis. Birmingham. ID Sept. 1812; FD 1813.
   BAYONETS. Delivered 100 sheet-iron musket bayonet scabbards for trials Sept. 1813.

DEAKIN, Mary. Birmingham. ID Jan. 1823 – FD Jan. 1824.
   BARRELS and LOCKS, Pattern 1823 Infantry Rifle.

DEAKIN, Samuel. Birmingham. ID Dec. 1823; FD Feb. 1828.
BARRELS and LOCKS, Pattern 1823 Infantry Rifle; RAMMERS for Pattern 1798 Heavy Dragoon Carbines and Pistols for RSAF Enfield in 1828.

DEAKIN, William. Birmingham. ID Mar. 1805 – Jan. 1810; Sept. – FD Nov. 1823.
BARRELS; BARRELS and LOCKS, Pattern 1823 Infantry Rifle in 1823.

DEAKIN & SON, Francis. Birmingham. ID Oct. 1823; FD June 1828.
BARRELS and LOCKS, Pattern 1823 Infantry Rifle, and RAMMERS for RSAF Enfield in 1828.

DEAKIN & SON, William. Birmingham. ID Jan. 1810 – FD Feb. 1818.
BARRELS. Dismissed as contractor for bribing a Viewer, but still recorded as delivering in Feb. 1818.

DEELY, Thomas. Birmingham. ID Mar. 1806 – FD Apr. 1809.
BREECH PINS [PLUGS]

DENNISON, John. London. ID Nov. 1718 – FD Dec. 1720.
STOCKER & SETTER UP of Pattern 1718 muskets.

DICK, Walter. London. ID Dec. 1796 – FD June 1811.
COMPLETE ARMS, LOCKS and Cannon Locks. Gunsmith in the Small Gun Office in the Tower, did much repairing of locks, worked with Joseph Sherwood in supplying Short Land muskets during mid-1790s.

DOLEP, Andrew. London. ID Dec. 1703 – FD June 1711.
COMPLETE ARMS.

DOUGHTY, Thomas. London. Mar. 1696 only. Mary, for Thomas Mar. 1697.
COMPLETE ARMS.

DREW, John. London. ID Apr. 1715 – FD Aug. 1720.
STOCKER & SETTER UP, latterly of Pattern 1718 muskets.

DUCE, J. London. July 1838.
LOCKS, Pattern 1823 Infantry Rifle.

DUCE, John. Birmingham. ID Mar. 1806 – FD Feb. 1818.
LOCKS. It is not clear whether this is the same John Duce who was Foreman of lockmakers at the Lewisham manufactory.

DUCE, Mary & John. London. Apr. 1825; ID Oct. 1826; FD Feb. 1828.
LOCKS, Pattern 1823 Infantry Rifle 1825-6, for Pattern 1798 Heavy Dragoon Carbines and Pistols for RSAF Enfield in 1828.

DYER. Richard. London. ID Oct. 1688 – FD Sept. 1715.
COMPLETE ARMS to Feb. 1712. BARRELS and LOCKS (Nov. 1699 only). STOCKER & SETTER UP from Mar. 1715.

DYMOND (DIAMOND), Charles. London. ID Sept. 1692 – FD Nov. 1713.
BARRELS and LOCKS (including Nov. 1699); COMPLETE ARMS, including Government-gift New York Indian fusils in Feb. 1710 and muskets for Irish service in Nov. 1713, along with barrels for the latter.

E

EAMES (EMES), Josiah. Birmingham. ID Mar. 1806 – FD Dec. 1816.
RAMRODS.

EBBUTT, Lancelot. London. ID Dec. 1704 – FD Mar. 1716.
COMPLETE ARMS to June 1711 including Government-gift New York Indian fusils in Feb. 1710. STOCKER & SETTER UP from Mar. 1715.

EDGE, Thomas. Wednesbury. ID Jan. 1799 – FD July 1801.
BARRELS, India Pattern musket.

EDGE, Richard (senior & junior). Wednesbury. ID Aug. 1757 – FD Jan. 1760.
BARRELS and LOCKS, RAMRODS from 1762.

EDGE & SON, Richard. Wednesbury. ID Feb. 1760 – FD May 1774.
BARRELS and LOCKS, RAMMERS from 1762.

EGG, Durs. London. ID May 1783 – May 1786; June 1793 – Mar. 1818; Feb.1821; Jan. 1824 – FD Mar. 1827. ROUGH STOCKED & SET UP, COMPLETE ARMS. Breech-loading Crespi-system carbines in 1780s. General arms from 1793. Supplied brass-barrelled pistols for the Duke de Bouillon's espionage and privateering force on the Island of Jersey, 1796. Pistols and Infantry Rifles in 1821, Pattern 1823 Infantry Rifles only after 1821.

EGG, Joseph. London. ID Oct. 1816; Dec. 1820; 1821; Feb. 1824 – May 1827; 1835 – FD 1837. ROUGH STOCKED & SET UP, carbines; Infantry Rifles in 1820 and 1824 – 37, pistols in 1821. 16" Cavalry Carbines for the Queen of Spain in 1835 – 6, Infantry Rifles for the Shah of Persia in 1836, carbines and pistols in 1836-7.

EVERARD, William. London. ID Nov. 1718 – FD Dec. 1719.
STOCKER & SETTER UP of Pattern 1718 arms.

F

FALKNER & CO., Edward. Birmingham. ID Feb. 1778 – FD June 1779.
BARRELS, LOCKS, BAYONETS and RAMRODS.

FARLOW, John. Birmingham. ID June 1719 - Mar. 1720; Dec. 1728 - Nov. 1730; FD Sept. 1737.
LOCKS, musket. Only one delivery in 1737.

FARMER, James. Birmingham. ID May 1741 – Aug. 1751; Aug. 1757 – FD May 1759.
LOCKS (1741 – 1751), BARRELS and COMPLETE ARMS. Locks and barrels only from 1757. Complete arms Mar. 1746 – Sept. 1748. Successor to his father Joseph.

FARMER, Joseph. Birmingham. ID Nov. 1708 – Oct. 1710; Mar. – Dec.1718; June 1723 – May 1732; FD May 1741.
COMPLETE ARMS to Oct. 1710, June 1723 - Jan. 1728; otherwise BARRELS and LOCKS.

FARMER & GALTON. Birmingham. ID Dec. 1756 – FD May 1774.
LOCKS and BARRELS. Partnership of James Farmer and Samuel Galton.

FEARNLEY, Ann. London. ID Sept. 1813 – Mar. 1818; 1820 – 1821; FD Mar. 1824.
ROUGH STOCKER, SETTER UP and COMPLETE ARMS. Widow of Robert whom she succeeded. Supplied 1813 Indian arms. Pistols in 1821, Infantry Rifles only in 1820 and 1824.

FEARNLEY, Robert. London. ID Feb. 1804 – FD Aug. 1813.
ROUGH STOCKER, SETTER UP and COMPLETE ARMS. Had done repair work for the Board prior to the above date.

FINCH, John. London. ID Oct. 1688 – FD July 1690. Widow Ann for, to Feb. 1691.
COMPLETE ARMS.

FISHER, George junior. London. ID Oct. 1688 – FD Jan. 1693.
COMPLETE ARMS. Furbisher to the Small Gun Office from 1689, died 1694 and succeeded by Henry Crips.

FISHER, Mary. London. ID Nov. 1693 – FD Sept. 1705.
COMPLETE ARMS. BARRELS and LOCKS (Nov. 1699 only). Took over husband George's contracts. Margaret paid for Mary Sept. 1705.

FITCHETT, William. London. June 1719 only.
SETTER UP of one delivery of Pattern 1718 muskets.

FLETCHER, Thomas. Birmingham. ID Mar. 1806 – FD Feb. 1818.
LOCKS.

FLETCHER & CO. Birmingham. Oct. 1823 only.
LOCKS, Pattern 1823 Infantry Rifle.

FORT, Mary. London. ID Jan. 1709; Mar. – FD Sept. 1715.
SETTER UP.

FORT, Thomas. London. ID Feb. 1691 – FD Dec. 1711.
COMPLETE ARMS to Mar. 1711 and BARRELS for Irish service muskets in Dec. 1711.

FRANKLAND, Richard. London. Mar. 1720 only.
ROUGH STOCKER Pattern 1718 Sea Service muskets.

FREEMAN, James. London. ID Sept. 1723; FD Jan. 1728.
COMPLETE ARMS: regimental muskets to Maj. Gen. Sabine's Prince of Wales's Own Regiment of Welsh Fusiliers in Sept. 1723. Another set to Lord Cadogan's Grenadier Guards in Nov. 1727. A final regimental set, possibly for two, to unnamed unit(s) in Jan. 1728.

FREEMAN, Samuel. Location unknown. ID July 1756 – FD Sept. 1781.
WOODEN RAMRODS.

G

GALE, Charles. London. ID Oct. 1688 – FD Dec. 1701.
COMPLETE ARMS.

GALE, Mary. London. ID June 1702 – FD June 1706.
COMPLETE ARMS. Executrix for Charles.

GALTON, Samuel. Birmingham. ID Sept. 1756 – FD Aug. 1759   (then see **FARMER & GALTON**).   BARRELS and LOCKS.

GALTON & SON, Samuel. Birmingham. ID May 1775 – Apr. 1787; Dec.1793 – FD Jan. 1796.
BARRELS, LOCKS, RAMMERS. One of the four contractors for COMPLETE Pattern 1776 muzzle loading, and Ferguson rifles in 1776. Added BAYONETS from Mar. 1778.

GALTON, Samuel junior. Birmingham. ID Jan. 1796 – FD Sept. 1819.
BARRELS, LOCKS, RAMMERS, ROUGH STOCKED & SET UP and COMPLETE ARMS including 1813 Indian arms.   Gave up supplying barrels, locks and rammers from Jan. 1814, and only set up arms.

GANDON, Peter (junior). London. ID May 1743 – May 1749; FD Dec.1756.
SETTER UP. One of eight new firms brought in, beginning in 1742, to increase wartime output. One delivery only in 1756.

GARDINER, Thomas (senior). London. ID Oct. 1688 – FD Nov. 1693.
COMPLETE ARMS. Mary for Thomas Apr. 1695.   Storekeeper in the Small Gun Office in the Tower, 6 June 1691 to his death in 1695.

GILBERT, Ann. London. ID Feb. 1691 – FD Mar. 1696.
COMPLETE ARMS.

GILL, Elizabeth. Birmingham. ID Mar. 1817 – FD May 1819.
ROUGH STOCKED & SET UP, BARRELS and BAYONETS. Widow of John, took over his contracts. **& Sons**, Nov. 1823 only: Barrels and locks, Pattern 1823 Infantry Rifle.

GILL, Thomas. Birmingham & London. ID June 1778 – Apr. 1781; Dec.1793 – FD Aug. 1801.
TOOLS and BAYONETS 1778-81.   BARRELS, LOCKS, BAYONETS. ROUGH STOCKED & SET UP and COMPLETE ARMS from 1793.

GILL, Thomas & James. Birmingham. Sept. 1801 only.
BARRELS, LOCKS, COMPLETE ARMS.

GILL, Thomas, James & John. Birmingham. Nov. 1801 only.
COMPLETE ARMS: Heavy Dragoon Carbines and Pistols.

GILL, John. Birmingham. ID Dec. 1801 – FD Mar. 1802.
BARRELS, LOCKS, COMPLETE ARMS.

GILL & CO., John. Birmingham. ID Feb. 1805 – FD Feb. 1817.
BARRELS, LOCKS, BAYONETS, ROUGH STOCKED & SET UP and COMPLETE ARMS including 1813 Indian arms.   Also BARREL RIFLER, *circa* 1812.

GILL & HADLEY. Birmingham. Apr. 1807 only.
Component suppliers, not specified but probably bayonets, barrels and ramrods.

GLASCOTT, Mary & G. M., London. ID Aug. 1823 – FD June 1826.
BRASS GUN FURNITURE, Pattern 1823 Infantry Rifle, and for Pattern 1798 Heavy Dragoon Carbines and Light Infantry Serjeant's Carbines for RSAF Enfield in 1826.

GLASCOTT BROTHERS. London. Apr. 1838 only.
BRASS GUN FURNITURE, infantry rifle.

GLUVIAS, Hewit. London. ID Sept. 1743 – FD Nov. 1744.
SETTER UP of Long Land muskets.   One of eight firms brought in and beginning in 1742 to increase output.

GODWARD (GODDARD), Edward. London. ID Nov. 1691 – FD Sept. 1709.
 COMPLETE ARMS. BARRELS and LOCKS (Nov. 1699 only).

GOFF (GOFFE), Benjamin. London. ID Nov. 1692 – Mar. 1696; Mar. 1706 – FD Feb. 1711.
 COMPLETE ARMS.

GOFF, Daniel. London. ID Feb. 1782 – Feb. 1783; FD Oct. 1785.
 SETTER UP to 1783. Three deliveries of carbine BARRELS in 1785.

GOODBY, James. London. ID Dec. 1701 – FD June 1716.
 COMPLETE ARMS to Sept. 1712. BARRELS for Irish service muskets in Dec. 1713. Supplied Government-gift Indian fusils in Feb. 1716. STOCKER & SETTER UP from Mar. 1715.

GREAVES, Joseph. Birmingham. ID Sept. 1812 – Dec. 1816; Jan. 1823 – FD 1824.
 BARRELS, India Pattern musket; Infantry Rifle in 1823-4.

GREEN, John. London. ID June 1710 – FD June 1711.
 LOCKS.

GREEN, Mary. London. July 1728 only.
 WOODEN RAMMERS.

GREEN, Thomas. London. ID Mar. 1697 – Apr. 1715; Mar. – FD July 1728.
 COMPLETE ARMS to June 1711, including Government-gift New York Indian fusils in Feb. 1710. LOCKS for Irish service muskets in Nov. 1713. STOCKER & SETTER UP Mar. 1715 – Dec.1715; Mar.– May 1728. WOODEN RAMMERS 1727.

GREEN, Thomas & Hezekiah. London. ID Jan. 1780 – Dec. 1782; 1786-1788; May – FD Oct. 1794. LOCKS and BARRELS until 1788, then BARRELS only. Small deliveries of carbine barrels in 1786 and 1788. Thomas Price joined firm in 1794, see Green & Price and Theodore & Philomon Price.

GREEN, PRICE & PRICE. Birmingham. ID 1804 – FD Feb. 1805.
 BARRELS.

GREEN & PRICE. Birmingham. ID Oct. 1794 – FD Mar. 1804.
 BARRELS.

GREGORY, Richard. London. ID Apr. 1704 – FD Nov. 1707.
 COMPLETE ARMS. Also Lock Viewer in The Tower from Sept. 1704.

GRICE, John & William. Birmingham. Mar. 1742 only.
 COMPLETE ARMS. Land Service Pistols and Carbines, one contract only.

GRICE, Joseph (I). Birmingham. ID Oct. 1762 – FD July 1771.
 BARRELS and LOCKS. Two men of this name, uncle and nephew. The nephew informed the Board in Nov. 1773 that his uncle was dead and that he wished to succeed to his contracts, which was approved.

GRICE, Joseph (II). Birmingham. ID Dec. 1773 – Mar. 1783; Jan.1794 – FD June 1804.
 BARRELS, LOCKS, and from Apr. 1797 ROUGH STOCKED & SET UP and COMPLETE ARMS. Supplied pattern Infantry Rifles in April 1800.

GRICE & EDGE. Birmingham. ID July 1756 – FD Dec. 1759.
 BARRELS and LOCKS. Partnership of William Grice and Richard Edge.

GRICE & MORRIS. Birmingham. ID May 1797 – FD Aug. 1800.
 LOCKS and COMPLETE ARMS. Partnership of widow Grice and Henry Morris. See Morris & Grice.

GRICE & SON, William. Birmingham. ID Apr. 1760 – Nov. 1782; Feb. 1789 – FD Mar. 1790.
 LOCKS, RAMMERS and some BARRELS. Major supplier of locks, and lock repair work to 1782. Mention in May 1772 of Grice senior. One of the four contractors for COMPLETE Pattern 1776 muzzle-loading and Ferguson rifles in 1776. Senior died 24 July 1790.

GRIGG, Abigail. Birmingham. ID Jan. 1814 – FD Dec. 1816.
 RAMRODS. Succeeded John.

GRIGG, John. Birmingham. ID Apr. 1812 – FD Dec. 1813.
 RAMRODS.

GROOME, Collins. London. ID Oct. 1688 – FD July 1690.
 COMPLETE ARMS.

GROOME, Margaret. London ID July 1690 – FD Apr. 1695.
 COMPLETE ARMS. Widow of Collins and received payments for arms supplied in his name until Jan. 1693, then in her own name to Apr. 1695, when payment for a second delivery of the same date is made to Mary for Margaret.

GROOME, Richard. London. ID Oct. 1688 – FD July 1690.
 COMPLETE ARMS. Succeeded by Collins.

GUNBY, John. Birmingham. Dec. 1823 only.
 BARRELS and LOCKS, Pattern 1823 Infantry Rifle. Had earlier been in partnership with Henry Osborn.

GYDE, John. London. ID 1786 – FD June 1822.
 ENGRAVER.

H

HADLEY, Thomas. Birmingham. ID June 1776 – Aug. 1781; May 1804 – Mar. 1818; Oct. – Nov. 1823; FD Mar. 1828. BARRELS, LOCKS, BAYONETS, SWORDS and RAMMERS. BAYONETS 1779 – 81, and from 1804 on. This may be two men of the same name: there is one mention of T.W. Hadley in Jan. 1815, supplying the same materials. Pattern 1823 Infantry Rifle barrels and locks only in 1823. Steel RAMMERS for Infantry Rifles for RSAF Enfield in 1828.

HADLEY, Henry. London. July 1757 only.
 SETTER UP.

HALFHIDE, George. London. ID Mar. 1719 – FD Nov. 1720.
 SETTER UP of Pattern 1718 muskets.

HALL, John. London. ID June 1706 – Nov. 1720; FD Apr. 1728.
 COMPLETE ARMS to June 1712. BARRELS for Irish service muskets Nov. 1713. STOCKER & SETTER UP from Feb. 1715 – Nov. 1720; one delivery in 1728.

HALL, Joseph. London. ID Aug. 1756 – FD Sept. 1757.
 SETTER UP.

HAMPTON, Thomas. Birmingham. ID Oct. 1812 – Jan. 1818; FD June 1825.
    BARRELS, ROUGH STOCKED & SET UP, COMPLETE ARMS, including 1813 Indian arms. Pattern 1823 Infantry Rifles only in 1825.

HARRIS, Joseph (I). Birmingham. ID Jan. 1757 – FD June 1762.
    BARRELS and LOCKS.

HARRIS, Joseph (II). Birmingham. ID Aug. 1812 – Feb. 1818; Apr.1823 – FD 1824.
    BARRELS. Pattern 1823 Infantry Rifle only 1823-4.

HARRIS, William. London. ID Feb. 1808 – FD Aug. 1811.
    BRASS GUN FURNITURE.

HARRIS & BARKER. Birmingham. ID Apr. 1762 – FD June 1771.
    BARRELS. Partnership of Joseph Harris and Matthias Barker. With the death of Harris, Barker formed a short-lived partnership with widow Jane Harris.

HARRISON, John. London. ID Sept 1778 – FD July 1779.
    COMPLETE ARMS. Became Harrison & Thompson.

HARRISON, Timothy. London. ID Dec. 1701 – FD Mar. 1716.
    COMPLETE ARMS to Mar. 1711. STOCKER & SETTER UP from Mar. 1715.

HARRISON & THOMPSON. London. ID Sept. 1778 – Mar. 1783; July 1793 – FD June 1805.
    ROUGH STOCKER, SETTER UP, LOCKS and COMPLETE ARMS. Partnership of Penelope Harrison (widow of John) and James Thompson.

HARTWELL, John. London. ID Oct. 1688 – FD June 1691.
    COMPLETE ARMS. Sent to Dublin early in 1689 and paid for work there Feb. 1689 to June 1691. Edward Hubbard received two payments for deliveries on behalf of Hartwell in July 1690 and February 1691.

HARTWELL, Susannah. London. ID June 1689 – FD July 1690.
    COMPLETE ARMS. Wife completing contracts of John.

HARTWELL & HASKINS. Birmingham. May 1775 only.
    LOCKS. Winding-up partnership of Abigail Haskins (widow of George) and Thomas Hartwell.

HARTWELL & MAYOR. London. ID Feb. 1755 – FD Dec. 1767.
    BRASS GUN FURNITURE. Partnership of Joseph Hartwell and Thomas Mayor. Virtual monopoly suppliers during their time. Mayor continues after Hartwell's death.

HARVEY, George & Samuel. Birmingham. ID Feb. 1779 – FD Apr. 1781.
    BAYONETS.

HARVEY, Robert. London. ID Dec. 1703 – FD Feb. 1705.
    COMPLETE ARMS, Pattern 1703 muskets.

HASKINS, George. Birmingham. ID June 1757 – Feb. 1759; Apr. 1767 – FD Dec. 1769.
    LOCKS. Succeeded by partnership with George Vernon until 1767, then under his own name to his death in 1775.

HASKINS & VERNON. Birmingham. ID Dec. 1758 – FD Dec. 1766.
    LOCKS.

HATCHER, Thomas. London. ID Dec. 1745 – FD Jan. 1752.
   COMPLETE ARMS. A gunsmith in his own right, appointed as Master Furbisher at the Small Gun Office in April 1750 five years after the death of Richard Wooldridge (although he may have filled the post earlier). Made several small groups of Land Pattern muskets, pistols, and the first group of carbines for Royal Artillery officers in 1750 and several succeeding small groups. In October 1751 he made a fusil for the Lieut. Gen. of the Ordnance. Took his son Thomas into partnership as Hatcher & Son 1769. Dead in May 1772. Replaced as Master Furbisher by William King in the same month.

HAWDEN, Flower. London. ID Jan. 1706 – FD May 1711.
   COMPLETE ARMS.

HAWGOOD, John. London. ID Dec. 1703 – FD Jan. 1708.
   BAYONETS.

HAWGOOD, Thomas. London. ID Dec. 1703 – FD Aug. 1711.
   BAYONETS.

HAWKINS, John senior. London. ID Oct. 1688 – FD Mar. 1716.
   COMPLETE ARMS to June 1711 including Government-gift New York Indian fusils in Feb. 1710; BARRELS and LOCKS for Irish service muskets Dec. 1713. STOCKER & SETTER UP from Dec. 1714.

HAWKINS, John junior. London. ID Feb. 1715; FD Sept. 1715.
   STOCKER & SETTER UP.

HAWKINS, Richard. London. ID Dec. 1701 – FD June 1708.
   COMPLETE ARMS.

HAWLEY, Joseph. London. ID Dec. 1703 – FD June 1705.
   COMPLETE ARMS.

HAWLEY, Thomas. London. ID Oct. 1688 – FD Mar. 1706.
   COMPLETE ARMS. BARRELS and LOCKS (Nov. 1699) and LOCKS (1706).

HAYNES, James. London. ID May 1706 – FD Sept. 1716.
   COMPLETE ARMS to May 1711. STOCKER & SETTER UP from Mar. 1715.

HEASLER, William. London. ID July 1692 – FD Sept. 1717.
   COMPLETE ARMS to June 1711 including Government-gift New York Indian fusils in Feb. 1710, muskets for Irish service and BARRELS for the same service in Nov. 1713. BARRELS and LOCKS (Nov. 1699 only). STOCKER & SETTER UP from Sept. 1715.

HEELY, James. Birmingham. ID Apr. 1815 – FD May 1815.
   BARRELS.

HEELY & CO., John. Birmingham. ID Feb. 1816 – Feb. 1818; FD Nov. 1823.
   BARRELS. Pattern 1823 Infantry Rifle only in 1823.

HEELY & CO., John & Samuel. Birmingham. ID Dec. 1794 – FD Dec. 1815.
   BARRELS. BAYONETS in 1803.

HENNEM, Jonathan. Lewisham, Kent. ID Nov. 1783 – FD Mar. 1805.
REFURBISHING, BARRELS, LOCKS, ROUGH STOCKER, SETTER UP, COMPLETE ARMS. Leaseholder of the Ordnance works at Armoury Mills on a 21 year lease approved Dec. 1784, chiefly involved with refurbishing edged weapons and pole arms. Supplied a screwless lock of his design for experimental arms in 1784, and supplied Short Land muskets sporadically from 1785 to 1793, then steadily 1793 – 1799, and under 150 Duke of Richmond's muskets in 1793.

HENSHAW, Thomas. London. ID Jan. 1757 – FD July 1757.
SETTER UP. Two deliveries.

HEPTINSTALL, William. London. July 1838 only.
ROUGH STOCKED & SET UP Pattern 1823 Infantry Rifles.

HILL, John. Birmingham. ID July 1812 – FD 1816.
BAYONETS.

HIRST, James. London. ID June 1777 – FD Oct. 1784.
SETTER UP. Partner of and successor to his father John.

HIRST, John. London. ID Feb. 1746 – FD Sept. 1776.
SETTER UP.   Held virtual monopoly of setting up all types of Ordnance small arms 1757 – 1775. Made the first breech-loading rifles tried by the Ordnance in 1762 and 1764. Took son James into partnership 15 Apr. 1774.   Repaired large quantities of barrels and refurbishing of arms. Dead by 30 Oct. 1776, son James continued the business.

HISCOCK, John. London. Oct. 1805 only.
COMPLETE ARMS. One delivery.

HODGSON, Elizabeth. London. ID June 1689 – FD Feb. 1691.
COMPLETE ARMS. Payments received for deliveries on behalf of Thomas.

HODGSON, Paul. London. ID July 1690 – FD Jan. 1693.
COMPLETE ARMS. Payments received on behalf of Elizabeth, then in own name 1691-93.

HODGSON, Thomas. London. ID Oct. 1688 – FD Jan. 1693.
COMPLETE ARMS.

HOLDEN, William. Birmingham. ID Jan. 1776 – Feb. 1783; Sept. 1786; Jan. 1788; Oct. 1794 – FD Mar. 1799.   BARRELS and LOCKS (1776-1780; 1782). Barrels only from 1783.

HOLLIER, Thomas. Lewisham, Kent. ID Sept. 1716 – FD Feb. 1753.
IRON and BRASS GUN FURNITURE, BAYONETS, IRON RAMRODS, SWORDS, HANGERS and POLE ARMS.   Leaseholder of the Armoury Mills.   Supplied iron and brass gun furniture, on a virtual monopoly basis from 1728 to 1748; also repaired and refurbished edged weapons, bayonets, and pole arms.   Initials TH with or without a Broad Arrow are often found cast into the inner surface of brass furniture.   Dead by May 1754. Succeeded at Armoury Mills by Richard Hornbuckle.

HOLLIS, Richard & William. Birmingham. ID July 1812 – Jan. 1818; July 1825; 1836 – FD 1837. ROUGH STOCKED & SET UP, COMPLETE ARMS including 1813 Indian arms. Pattern 1823 Infantry Rifles only in 1825. Carbines and pistols in 1836-7.

HORNBUCKLE, Richard. Lewisham, Kent. ID May 1756 – FD 1784.
  REFURBISHING; BAYONETS from 1778. Leaseholder of the Armoury Mills for 31 years from May 1754 after Thomas Hollier's death. Took possession in Dec. 1754. Mostly re-furbishing work. Supplied musket BAYONETS from June 1778 to Nov. 1780. Dead by June 1784 when it was noted that his widow continued the work. Succeeded by Jonathan Hennem.

HORSLEY, John. Birmingham. ID June 1778 – FD May 1780.
  BAYONETS, musket.

HOW, Richard. London. ID Oct. 1756- FD Sept. 1758.
  SETTER UP.

HOWELL, William. Birmingham. ID Oct. 1823 – FD July 1825.
  LOCKS, Pattern 1823 Infantry Rifle.

HUGGINS, William. Birmingham. ID Apr. 1704 – FD Dec. 1730.
  LOCKS (May 1704–June 1709), COMPLETE ARMS (June 1706–June 1711), IRON and BRASS GUN FURNITURE (1727-30), BAYONETS (1727-30) and IRON RAMMERS (Mar. 1727).

HUMPHRYS (HUMPHRIES), Veritas. London. ID Jan. 1778 – FD Nov.1781.
  SETTER UP, Land muskets and a few Sea Service muskets.

HUNT, Joseph. London. ID Feb. 1778 – FD May 1781.
  ROUGH STOCKER & SETTER UP, Short Land Pattern muskets.

HUNT, Thomas. London. ID Mar. 1708 – FD Mar. 1712.
  COMPLETE ARMS, LOCKS (1709), BARRELS (1712).

J

JACKSON, Christopher. London. ID Mar. 1709 – FD June 1715.
  COMPLETE ARMS. STOCKER & SETTER UP from Mar. 1715.

JAMES, Henry. Birmingham. ID June 1815 – FD Dec. 1816.
  BARRELS.

JOHNSON, Christopher. London. ID Jan. 1708 – FD Sept. 1716.
  COMPLETE ARMS to Sept. 1712 and muskets for Irish service Dec. 1713. STOCKER & SETTER UP from June 1715.

JOHNSON, Francis. Birmingham. Nov.– Dec. 1823 only.
  BARRELS and LOCKS, Pattern 1823 Infantry Rifle.

JOHNSON, John. London. ID Mar. 1696 – FD June 1720.
  COMPLETE ARMS to Dec. 1713 including muskets for Irish service in Dec. 1713. STOCKER & SETTER UP from Feb. 1715.

JOHNSON, Margaret. London. ID Mar. 1718 – Nov. 1720; FD Mar. 1729.
  ROUGH STOCKER & SETTER UP.

JORDAN, Edward. Birmingham. ID Feb. 1733 – FD July 1758.
  Principal supplier of BARRELS and LOCKS (Oct. 1738 – Mar. 1748; Dec. 1755 – July 1758). COMPLETE ARMS Dec. 1746 – June 1748. Died 17 May 1758.

JORDAN, Thomas. Birmingham. ID Apr. 1744; Mar. 1757 – FD July1762.
   BARRELS, LOCKS, RAMMERS and BAYONETS. Son of Edward.

JORDAN & FARMER. Birmingham. ID July 1742 – FD June 1746.
   LOCKS and COMPLETE ARMS. Partnership of Edward Jordan and James Farmer.

K

KELSO, Isaac. London. ID May 1756 – FD Sept. 1757.
   ROUGH STOCKER.

KETLAND, Elizabeth. Birmingham. ID Mar. 1816 –May 1819; FD Nov.1823..
   BAYONETS, RAMMERS, ROUGH STOCKED & SET UP, chiefly New Land Pattern muskets, and a few carbines. Pattern 1823 Infantry Rifle LOCKS only in 1823.

KETLAND, Thomas. Birmingham. Oct. 1823 only.
   LOCKS, Pattern 1823 Infantry Rifle.

KETLAND, William. Birmingham. Apr. 1804 only.
   BARRELS, BAYONETS and COMPLETE ARMS.   Died Nov. 1804.

KETLAND, WALKER & CO. Birmingham. ID Sept. 1808 – Sept. 1819; FD Oct. 1823.
   ROUGH STOCKED & SET UP, LOCKS, COMPLETE ARMS including 1813 Indian arms. John Adams was taken into the company. Under this firm style became the largest single supplier of small arms to the Ordnance during the Napoleonic Wars, and the largest single supplier of India Pattern muskets. Pattern 1823 Infantry Rifle locks only in 1823.

KETLAND & ALLPORT. Birmingham. ID Mar. 1804 – FD May 1819.
   ROUGH STOCKED & SET UP, COMPLETE ARMS including 1813 Indian arms, LOCKS, BAYONETS and RAMMERS. Second largest producer of India Pattern muskets. William Allport starts as a separate contractor Feb. 1816, and also Elizabeth Ketland. William Ketland takes over, but still under the K&A firm style at the Board's insistence.

KETLAND & WALKER. Birmingham. ID Jan. 1796 – FD Aug. 1808.
   ROUGH STOCKED & SET UP, LOCKS,COMPLETE ARMS, largest single supplier in the 1804 – 1815 period. Became Ketland, Walker and Co. Sept. 1808.

KETLAND & CO., W. Birmingham. July 1825 only.
   ROUGH STOCKED & SET UP, Pattern 1823 Infantry Rifles.

KIPLING, Charles. London. ID July 1690 – FD June 1715.
   COMPLETE ARMS to Mar. 1712 including Government-gift New York Indian fusils in Feb. 1710. BARRELS and LOCKS (Nov. 1699) and for Irish service muskets Mar. 1714. STOCKER & SETTER UP from Mar. 1715.

KIPLING, Richard. London. ID Dec. 1701 – FD Apr. 1715.
   COMPLETE ARMS to July 1711.   LOCKS (1707). STOCKER & SETTER UP from Apr. 1715, one delivery.

KIRBY, Henry. London. ID Apr. 1689 – FD Dec. 1706.
   COMPLETE ARMS.

KIRKHAM, Henry. London. ID June 1716 – Mar. 1721; FD Apr. 1728.
STOCKER & SETTER UP, including Government-gift Indian fusils in July 1716.

KNUBLEY, John. London. ID May 1793 – FD Nov. 1794; his widow made one delivery in April 1795. ROUGH STOCKED & SET UP and COMPLETE ARMS. Also supplied commercial military-pattern arms purchased in the open market.

L

LACY & CO. London. July 1838 only.
ROUGH STOCKED & SET UP Pattern 1823 Infantry Rifles.

LACY & REYNOLDS. London. ID Aug. 1835; 1836; FD 1837.
ROUGH STOCKED & SET UP carbines, pistols and Infantry Rifles. 16" Cavalry Carbines for the Queen of Spain in 1835; Pattern 1823 Infantry Rifles for the Shah of Persia in 1836, and pistols and carbines in 1836–7.

LAMPHEERE, Richard. London. Apr. 1728 only.
SETTER UP.

LAWRENCE, Margaret. London. Sept. 1715 only.
STOCKER & SETTER UP.

LAWRENCE, Richard. London. ID June 1689 – FD May 1711.
COMPLETE ARMS and BARRELS (1707).

LEIGH, James Brooks. London. July 1838 only.
ROUGH STOCKED & SET UP Pattern 1823 Infantry Rifles.

LEONARD, John. Lichfield, Staffs. ID Aug. 1780; FD 1782.
LOCKS.

LEONARD, William. Birmingham. ID Mar. 1806 – Feb. 1818; FD Oct.1823 .
LOCKS. Pattern 1823 Infantry Rifle only in 1823.

LIMBREY, Matthew. London. ID Aug. 1794 – FD Sept. 1797.
BARRELS, LOCKS, ROUGH STOCKED & SET UP and COMPLETE ARMS. Executors continued deliveries until Feb. 1801.

LOADER, Edward. London. ID Dec. 1707 – June 1712; FD Dec. 1717.
COMPLETE ARMS to June 1712. STOCKER & SETTER UP from Mar. 1715.

LOADER, Mary. London. ID Sept. 1719 – FD Dec. 1720.
STOCKER & SETTER UP of Pattern 1718 arms. Widow of Edward.

LOADER, Richard. London. ID June 1689 – FD Sept. 1715.
COMPLETE ARMS to Mar. 1711 including Government-gift New York Indian fusils in Feb. 1710. BARRELS and LOCKS (Nov. 1699 only). STOCKER & SETTER UP from Dec. 1714.

LODER, Joseph. London. ID Apr. 1756; June 1769 – FD Oct. 1785.
ROUGH STOCKER. One delivery of carbines in 1756; no other deliveries until June 1769 from which time he becomes, with James Waller, one of the principal rough stockers during this period.

LOOKER, Robert. London. ID Oct. 1688 – Oct. 1689; Mar. 1715 – FD Sept. 1718.
 COMPLETE ARMS. STOCKER & SETTER UP from 1715 including Government-gift New York Indian fusils in July 1716, also cleaned and repaired some of those recovered after shipwreck.

LOWNDES, Thomas. Birmingham. ID Feb. 1813 – FD Jan. 1818.
 ROUGH STOCKED & SET UP, COMPLETE ARMS including 1813 Indian arms.

LOXHAM, William & Edward. London. ID July 1754 – FD July 1780.
 BAYONETS and SWORDS. Uncle and nephew partnership, had a virtual monopoly of the bayonet supply until 1778. The uncle died in July 1780 and nephew Edward declined to continue his contracts.

LOYD, Evan. London. ID June 1716 – Mar. 1721; FD Apr. 1728.
 STOCKER & SETTER UP. One delivery in 1728.

LUTHER, Stephen. London. ID Mar. 1709 – FD June 1709.
 COMPLETE ARMS, two deliveries.

M

MAKIN, James. Birmingham. ID May 1795 – FD Oct. 1814.
 BAYONETS, India Pattern musket.

MANTON, Charles. London. ID 1829 – FD 1840.
 Master Furbisher, appointed on the death of Jonathan Bellis in 1829. An active designer of arms and accessories, he was instrumental in getting his pattern of cavalry carbine adopted to replace the Nock-lock and Office-lock Heavy Dragoon carbines from the 1790s, over strong objections from commanding officers of cavalry regiments. The design proved both a failure in itself and anachronistic, being replaced a few years later by a percussion carbine of larger bore and longer barrel. His design of nipple key was accepted by the Board in 1835. He produced a number of pattern arms during his short tenure.

MARWOOD, Ann. London. ID Jan. 1811 - May 1818; 1820 – 1821; Feb.1824; FD Mar. 1825.
 ROUGH STOCKER, SETTER UP, COMPLETE ARMS. Successor to her late husband William. Supplied several hundred New Land Pattern musket LOCKS from Apr. 1812 to Dec. 1814. Pattern 1823 Infantry Rifles only after 1821.

MARWOOD, William. London. ID 19 June 1798; FD 31 Dec. 1810.
 BARRELS, LOCKS, ROUGH STOCKER, SETTER UP, COMPLETE ARMS. Supplied several hundred New Land Pattern musket LOCKS from Sept. 1806. Died Dec. 1810, succeeded by widow Ann.

MAYO, John. London. ID Dec. 1701 – FD June 1720.
 COMPLETE ARMS to Sept. 1710, LOCKS (1708, 1719–1720), and did Sea Service refurbishing work.

MAYOR, Jane. London. ID July 1768 – FD Aug. 1795.
 BRASS GUN FURNITURE, principal supplier. Succeeded late husband Thomas. Succeeded by son Joseph in Sept. 1797.

MAYOR, Joseph. London. ID Sept. 1802 – Jun. 1816; FD Feb. 1819.
 BRASS GUN FURNITURE. Principal supplier. Succeeded his mother, Jane.

MAYOR, Thomas. London. ID June 1761 – FD Dec. 1767.
BRASS GUN FURNITURE. Originally partner with Joseph Hartwell. Succeeded by widow Jane.

MEAKIN, Nathaniel. London. ID May 1745 – FD Dec. 1755.
WOOD RAMMERS. Also many workshop and armourer's tools.

MEMORY, Michael. London. ID June 1756 – Aug. 1758; Oct. 1778 – Mar. 1783; May – FD July 1793. ROUGH STOCKER & SETTER UP, also COMPLETE Sea Service muskets (Feb. 1779–Apr. 1780). Became Memory & Wright.

MEMORY & WRIGHT. London. ID Aug. 1793 – FD Aug. 1802.
ROUGH STOCKER & SETTER UP. Also supplied non-regulation arms purchased in the open market. Partnership of Michael Memory and Robert Wright, who succeeded in his own name.

MEREDITH & MOXHAM. Birmingham. Nov. 1823 only.
BARRELS, Pattern 1823 Infantry Rifle.

MILLS, William. London. Apr. 1839 only.
ROUGH STOCKED & SET UP Pattern 1823 Infantry Rifles.

MILLS & SON, William. London. ID Aug. 1836; FD 1837.
ROUGH STOCKED & SET UP. Pattern 1823 Infantry Rifles for the Shah of Persia in 1836, and pistols in 1836–7.

MIST, Phillip. Birmingham. ID Aug. 1703 – June 1711.
BAYONETS.

MOLE, John & Robert. Birmingham. ID Aug. 1836; FD May 1838.
BAYONETS, Pattern 1823 Infantry Rifle 'Hand Bayonets'.

MOLINEUX, Richard. Birmingham. May 1759 only.
BARRELS, LOCKS and STEEL RAMMERS. One delivery.

MOORE, Daniel. London. ID Aug. 1778 – Dec. 1779; FD Aug. 1780.
COMPLETE ARMS, Sea Service muskets and a small number of Short Land muskets in Aug. 1780 only.

MOORE, Thomas. London. ID Mar. 1700 – FD Mar. 1711.
COMPLETE ARMS, LOCKS (1707, 1709).

MOORE, William. London. ID Sept 1689 – FD June 1720.
COMPLETE ARMS to May 1711 including Government-gift New York Indian fusils in Oct. 1702. BARRELS and LOCKS (Nov. 1699 only). STOCKER & SETTER UP from Feb. 1715.

MORRIS, Anthony. London. ID Sept. 1692 – FD Mar. 1696.
COMPLETE ARMS.

MORRIS, Henry. Birmingham. ID Apr. 1804 – FD Jan. 1810.
BARRELS, LOCKS, BAYONETS (in 1804), ROUGH STOCKED & SET UP, COMPLETE ARMS. Succeeded by Morris & Grice.

MORRIS, Isaac. London. ID Dec. 1701 – FD June 1721.
COMPLETE ARMS to Jan. 1712. STOCKER & SETTER UP from Mar. 1715.

MORRIS & GRICE. Birmingham. ID Feb. 1810 – Sept. 1819; Oct. – FD Dec.1823.
BARRELS, RAMMERS, ROUGH STOCKED & SET UP and COMPLETE ARMS including 1813 Indian arms. Partnership of Henry Morris's widow and Joseph Grice the younger. Pattern 1823 Infantry Rifles only in 1823.

MOXHAM, Thomas. Birmingham. ID Sept. 1810 – Sept. 1819; July 1825; 1836; FD 1837.
BARRELS, LOCKS, ROUGH STOCKED & SET UP and COMPLETE ARMS including 1813 Indian arms. Pattern 1823 Infantry Rifles only in 1825. Carbines in 1836 and pistols in 1837.

MUNTZ & CO., P.F. Birmingham. ID July 1812 – Dec. 1816; Nov. 1823; FD Nov. 1825.
BARRELS. Pattern 1823 Infantry Rifle barrels only in 1823 and 1825.

N

NEAVE, Richard. London. Sept. 1703 only.
COMPLETE ARMS.

NEGUS, Ann. London. May 1802 only.
LOCKS.

NEGUS, Deborah. London. Oct. 1823 only.
LOCKS, Infantry Rifle.

NEGUS, James (I?). London. ID Feb. 1794 – FD Aug. 1802.
LOCKS. In partnership with Ezekiel Baker 1802 to 1805 but no record of supplying the Ordnance under this style.

NEGUS, James (II?). London. ID Mar. 1804 – FD Feb. 1818.
LOCKS. Possibly same as last.

NELSON, John. London. ID July 1690 – FD June 1702.
COMPLETE ARMS. BARRELS and LOCKS (Nov. 1699 only).

NEW, Henry & Elizabeth. Birmingham. ID Jan. 1817 – FD Jan. 1818.
COMPLETE ARMS. New Land Pattern muskets and Cadet carbines.

NEWMAN, Thomas. Birmingham. ID Apr. 1814 – FD Dec. 1816.
RAMRODS.

NEWTON, Edward. Grantham, Lincs. ID Aug. 1759 – FD July 1761.
COMPLETE ARMS. One delivery of 600 Marine or Militia muskets for which he provided the brass furniture, smallwork, barrels, locks and the setting up.

NICHOLSON, Ann. London. Sept. 1794 only.
SETTER UP. Widow of William.

NICHOLSON, Edward senior. London. ID Oct. 1688 – FD Sept. 1711.
COMPLETE ARMS including Government-gift New York Indian fusils in June 1700, Oct. 1702 and Feb. 1710. BARRELS and LOCKS (Nov. 1699 only).

NICHOLSON, Edward junior. London. ID Nov. 1706 – FD Nov. 1710.
COMPLETE ARMS.

NICHOLSON, William. London. ID Mar. – Nov. 1782; Apr. 1795; FD May 1796.
ROUGH STOCKER and SETTER UP. Two deliveries in 1782, one in 1795 and one in 1796.

NOCK, Henry. London. ID Sept 1775 – Oct.1781; June 1784 – Oct.1792; Feb. 1793 – FD Mar. 1804. LOCKS, ROUGH STOCKER, SETTER UP, COMPLETE ARMS, EXPERIMENTAL ARMS. Supplied locks, especially Extra Flat Carbine locks, and seven-barrelled guns only during the American War to Oct. 1781. Made experimental arms at the behest of the Master General of the Ordnance (the third Duke of Richmond), during the 1780s and early 1790s, and was, with Jonathan Hennem, the principal rough stocker and setter up of regulation arms (mostly carbines) during the decade between the end of the American War and the commencement of the French Revolutionary Wars. From 1793, in addition to the normal range of regulation muskets, carbines and pistols, Nock produced, as complete arms, both patterns of the Duke of Richmond's Muskets and later altered some of them, as well as Harcourt's Pattern 1794 Heavy Dragoon Carbines (1794, 1797); he supplied patterns of the components and gauges for the Heavy Dragoon Carbine and Pistol in 1796; supplied complete brass-barrelled pistols for the Duke de Bouillon's espionage and privateering force on the Island of Jersey, 1796; produced Pattern 1797 Heavy Dragoon Carbines and Pistols (with Nock screwless lock), and Pattern 1798 Heavy Dragoon Carbines and Pistols (with Office-pattern locks); and supplied the first production of the Duke of York's Hussar Carbines (1800). His executors [James Wilkinson] completed outstanding contracts between Dec. 1804 and June 1805.

NUTT, William. London. ID June 1689 – FD Mar. 1716.
COMPLETE ARMS to Mar. 1712 including Government-gift New York Indian fusils in Feb. 1710. BARRELS and LOCKS (Nov. 1699) and for Irish service muskets Dec. 1713. STOCKER & SETTER UP from Mar. 1715.

O

OLDFIELD, Henry. London. Mar. 1696 only.
COMPLETE ARMS.

OSBORN, Hannah. Birmingham. Feb. 1828 only.
STEEL RAMMERS for Pattern 1798 Heavy Dragoon Carbines and Pistols for RSAF Enfield.

OSBORN, Henry. Birmingham. ID Aug. 1797 – Dec. 1807; Oct – Nov.1823; Dec. 1825; Jan.- FD June 1826. BAYONETS, SWORDS. Responsible for the design of the Pattern 1800 Baker Rifle sword-bayonet, for which he supplied the patterns. Became Osborn & Gunby in 1808 and separated again after 1818. BARRELS and LOCKS for Pattern 1823 Infantry Rifles 1823 to 1826. Supplied STEEL RAMMERS for Pattern 1798 Heavy Dragoon Carbines for RSAF Enfield in June 1826.

OSBORN, Thomas. Birmingham. ID Mar. 1806 – FD Aug. 1810.
BARREL RIFLER.

OSBORN & GUNBY. Birmingham. ID Jan. 1808 – FD Feb. 1818.
BARRELS, LOCKS, BAYONETS, RAMMERS, SWORDS. Partnership of Henry Osborn and John Gunby.

OUGHTON, John. Birmingham. ID Dec. 1796; FD Mar. 1803.
BARRELS and BAYONETS. In Sept. 1812 supplied iron musket bayonet scabbards for trials. Succeeded by partnership of John & Craxhall Oughton.

OUGHTON, John & Craxhall. Birmingham. ID Apr. 1803 – 1815; Oct. 1823 – FD Apr. 1826.
BARRELS and BAYONETS. Barrels and LOCKS for Pattern 1823 Infantry Rifles only from 1823.

OUGHTON, Joseph (I). Birmingham. ID Dec. 1755 – FD Mar. 1759.
BARRELS and LOCKS. One of the principal suppliers of barrels during his period of operation.

OUGHTON, Joseph (II). Birmingham. ID June 1776 – Mar. 1783; Mar.1787; Jan. 1788; Jan. 1794 – FD Apr.1794. BARRELS and BAYONETS. Succeeded by Oughton & Parr.

OUGHTON & PARR. Birmingham. ID July 1794 – FD Dec. 1796.
BARRELS.

OUGHTON & SON, Joseph. Birmingham, ID Mar. 1759 – FD June 1776.
BARRELS and LOCKS.

P

PARDOE, Ambrose. Liège. May 1780 only.
BAYONETS. During his tenure of 46 weeks and ten days as Ordnance Inspector of the Short Land Muskets being made in Liège for the Board, Pardoe supplied in his own name, one order of 10,100 Liège-made socket bayonets for Short Land muskets, for which he was paid as a regular contractor. Whether these were marked in any way to distinguish them from others being supplied is not known at the time of writing. Pardoe replaced William King as Master Furbisher in early 1781. He superintended the manufacture of the Norfolk Rangers' carbines and pistol-carbines by the workforce in the Small Gun Office Jan. to Apr. 1783. He recommended his successor, Jonathan Bellis, for the post of Assistant to the Master Furbisher in July 1793, and his wife Elizabeth was referred to as his widow in June 1794.

PARKER, Robert. Lewisham, Kent. ID Dec. 1707; FD June 1712 ?
Leaseholder of Armoury Mills.

PARKER, William. London. ID Nov. 1804 – Mar. 1818; Feb. 1821; Feb.1824; Feb. 1827; 1835 – 1836; July 1838; FD 1839. ROUGH STOCKED & SET UP and COMPLETE ARMS. Infantry rifles and pistols in 1821, Pattern 1823 Infantry Rifles only 1824-7; 16" Cavalry Carbines for the Queen of Spain in 1835–6; carbines, pistols. infantry rifles for the Shah of Persia, and Irish-service carbines 1835–9.

PARKIN, Thomas. London. ID Aug. 1808 – FD Apr. 1810.
MERCHANT who delivered commercially produced arms of military pattern bought in the open market, including India Pattern muskets.

PARR, James. Liverpool. ID Dec. 1803 – FD May 1810.
MERCHANT who delivered commercially produced arms of military pattern bought in the open market; also Extra Service India Pattern muskets, 1809-10.

PARSONS, Benjamin. Birmingham. ID July 1812 – Feb. 1818; Oct. 1823; FD July 1825.
  BARRELS. Pattern 1823 Infantry Rifle only in 1820s.

PARSONS, Gad. Birmingham. ID May 1798 – FD Oct. 1813.
  BARRELS. Became & Son Nov. 1813.

PARSONS, John. Birmingham. ID July 1812 – Feb. 1818; FD Nov. 1823.
  BARRELS. Pattern 1823 Infantry Rifle only in 1823.

PARSONS, Phineas. Birmingham. ID Sept. 1813 – Feb. 1818; FD Oct. 1823.
  BARRELS. Pattern 1823 Infantry Rifle only in 1823.

PARSONS & SON, Gad. Birmingham. ID Nov. 1813 – Feb. 1818; FD Dec. 1823.
  BARRELS, Pattern 1823 Infantry Rifle only in 1823.

PARTRIDGE, John. Birmingham. ID Sept. 1823 – FD Aug. 1826.
  LOCKS. Pattern 1823 Infantry rifle.

PARTRIDGE, W. Birmingham. Apr. 1838 only.
  LOCKS. Pattern 1823 Infantry Rifle.

PATTERSON (PATERSEN), Daniel. London. ID July 1807; FD May 1809.
  COMPLETE ARMS, one delivery of India Pattern muskets 1807, one delivery of Extra Service muskets 1809. Normally employed only for cleaning and repairing arms.

PEAKE, John. London. ID July 1803 – Mar. 1818; FD Nov. 1820.
  ROUGH STOCKED and SET UP and COMPLETE ARMS. Previously foreman to Alexander Raby. Infantry Rifles only in 1820.

PEDDELL (PEDDALL), James. London. ID Oct. 1688 – FD Mar. 1716.
  COMPLETE ARMS to June 1711 including Government-gift New York Indian fusils in Oct. 1702 and Feb. 1710, and muskets for Irish service in Dec. 1715. BARRELS and LOCKS (Nov. 1699 only). STOCKER & SETTER UP from Feb. 1715.

PERRY, William. Birmingham. ID Aug. 1756 – FD Mar. 1757.
  LOCKS, Land Musket. A William Perry was paid for two wallpieces and three blunderbusses delivered to Gen. Cleaveland on Gen. Gage's orders for the use of the troops at Boston, in Oct. 1775.

PETERS, Michael. Birmingham. ID Mar. 1806 – Feb. 1818; FD Nov. 1823.
  LOCKS. Pattern 1823 Infantry Rifle only in 1823.

PETTYMAN, Joseph. London. Nov. 1706 only.
  COMPLETE ARMS, one delivery of Pattern 1703 muskets.

PHILLIPS, Francis. London. ID Apr. 1689 – FD Mar. 1720.
  COMPLETE ARMS to Mar. 1711 including Government-gift New York Indian fusils in Feb. 1710. Muskets and barrels for Irish service Dec. 1713. BARRELS and LOCKS (Nov. 1699 only). STOCKER & SETTER UP from Feb. 1715.

PHILLIPS, Thomas. London ID May 1708 – FD Mar. 1723.
  COMPLETE ARMS to Apr. 1711. STOCKER and SETTER UP from Mar. 1715.

PICKFATT, Charles. London. ID Apr. 1727 – June 1749; July 1756 – FD July. 1757.
  BARRELS (1731 only), SETTER UP. With Lewis and then James Barber, chief setters up from 1730 to 1742, and continuing as a major producer through 1749.

PICKFATT, Humphrey senior. London. ID Oct. 1688 – FD Mar. 1696.
COMPLETE ARMS. Died 1698.

PICKFATT, Humphrey junior. London. ID 1689 – FD Oct. 1720.
COMPLETE ARMS to June 1712 including Government-gift New York Indian fusils in Feb. 1710. BARRELS and LOCKS (Nov. 1699 only). Muskets and barrels for Irish service in Dec. 1713. STOCKER & SETTER UP from Feb. 1715.

PIKE, Christopher. London. ID Oct. 1688 – FD Apr. 1695.
COMPLETE ARMS. Elizabeth for Christopher, 26 Mar. 1696 only.

PIKE, Elizabeth. London. Nov. 1699 only.
BARRELS and LOCKS.

PITTS, Mordecai. London. ID Oct 1704 – FD Dec. 1709.
COMPLETE ARMS.

PLANT, Benjamin. Birmingham. ID Sept. 1812 – FD Feb. 1818.
BARRELS.

PLANT, Martha. Birmingham. Nov. 1823 only.
BARRELS, Pattern 1823 Infantry Rifle.

POLLARD, James. London. ID June 1702 – FD Sept. 1705.
COMPLETE ARMS, one delivery of LOCKS (1703).

PORTLOCK, John. Birmingham. ID Sept. 1805 – FD Mar. 1809.
BARRELS. Dismissed by the Board for concurrently running a private commercial barrel shop.

PORTLOCK, Thomas. Birmingham. ID May 1798 – FD 1815.
BARRELS.

POTTS, Thomas. London. ID Apr. 1833; 1835 – FD Apr. 1839.
ROUGH STOCKED & SET UP, chiefly pistols and carbines; Pattern 1823 Infantry Rifles in 1836 (Shah of Persia), 1838 and 1839. Largest supplier of 16" Cavalry Carbines for the Queen of Spain in 1835–6. Also bored, rifled, sighted, smoothed and browned Infantry Rifle barrels in 1838 and fitted their Hand Bayonets. Formerly partner with Martin Brander.

POWELL, Hugh. London. ID Feb. 1716 – FD May 1720.
GUNSTOCK MAKER and ROUGH STOCKER and SETTER UP.

POWELL, Widow Martha. London. ID Mar. 1714; FD June 1715.
COMPLETE ARMS: one delivery of Irish muskets in Mar. 1714. STOCKER & SETTER UP of one delivery of muskets in 1714 and one in 1715. Widow of Stephen.

POWELL, Richard. London. ID Dec. 1718 – FD Dec. 1719
ROUGH STOCKER of two deliveries of Pattern 1718 muskets.

POWELL, Stephen. London. ID Mar. 1696 – Mar. 1711; FD Mar. 1715.
COMPLETE ARMS. STOCKER & SETTER UP Mar. 1715 only.

POWELL, Thomas. London. ID Dec. 1701 – FD Sept. 1711.
COMPLETE ARMS.

PRATT, John. London. ID May 1777 – FD July 1781.
ROUGH STOCKER and SETTER UP, COMPLETE ARMS, BARRELS and BAYONETS (May 1779-1780). The most important single contractor during the 1777–81 period. Designed a straight-tapered and collared second rammer pipe adopted as a standard feature in 1778-9. Fully Ordnance-marked Short Land Pattern muskets with this pipe were not made before March, 1779. Muskets supplied complete by Pratt (and marked by him) may have had this second pipe of his design by 1778. Pratt had previously worked for the Small Gun Office as a Viewer. Pratt was dismissed as a contractor in Nov. 1780 for bribing a Viewer. A year later he was applying to get his job back, and his dead stock removed, but there is no indication that he was again employed by the Board.

PREDDEN, William. London. ID Feb. 1691 – FD June 1720.
COMPLETE ARMS to June 1711 including Government-gift New York Indian fusils in Feb. 1710. BARRELS and LOCKS (Nov. 1699 only). Muskets for Irish service in June 1714. STOCKER & SETTER UP from Feb. 1715.

PRESS, Edward. London. ID Dec. 1703 – FD Mar. 1705.
COMPLETE ARMS, Pattern 1703 muskets.

PRESS, Joan. London. ID Sept. 1719 – FD Apr. 1720.
STOCKER & SETTER UP of Pattern 1718 muskets.

PRICE, Theodore. Birmingham. ID Dec. 1823; FD Jan. 1826.
LOCKS, Pattern 1823 Infantry Rifle.

PRICE, Theodore & Philomon. Birmingham. ID Feb. 1805 – FD 1815.
BARRELS. Headed a consortium of Birmingham barrel workers with a barrel-rolling and cock-stamping factory using machinery designed by Theodore Price.

PRIEST, Ann. Birmingham. ID Apr. 1817 – Feb. 1818; FD Nov. 1823.
BARRELS. Widow continuing with late husband's contracts. Pattern 1823 Infantry Rifle only in 1823.

PRIEST, Joseph. Birmingham. ID Sept. 1805 – FD Mar. 1817.
BARRELS.

PRINGLE, John. London. ID 1716 – FD June 1721.
STOCKER & SETTER UP, including Government-gift New York Indian fusils in July 1716, but chiefly of Pattern 1718 muskets.

PRITCHETT, Richard Ellis. London. ID Sept. 1809; Feb. 1824 –May 1827;1835 –FD Jan. 1839.
ROUGH STOCKED & SET UP and COMPLETE ARMS. One delivery of India Pattern Extra Service muskets; did much repair and furbishing work for the Board. Pattern 1823 Infantry Rifles 1824-7; 16" Cavalry Carbines for the Queen of Spain in 1835; then carbines, pistols and one delivery of Infantry Rifles each in 1836 (Shah of Persia), 1838 and 1839.

PRITCHETT, Samuel. London. ID Dec. 1793 – May 1808; FD 1809.
BARRELS, LOCKS, ROUGH STOCKER & SETTER UP, and COMPLETE ARMS.

PROBIN, Thomas. London. ID Sept. 1708 – Jan. 1711; Nov. 1717; FD Jan. 1728.
COMPLETE ARMS. In Dec. 1727 he supplied muskets and bayonets to Col. Montagu's Foot, and another set to an unnamed regiment in 1728.

PYE, Humphrey. London. ID Oct. 1688 – FD Aug. 1694.
COMPLETE ARMS.

PYE, Thomas. London. ID Nov. 1693 – FD Dec. 1706.
  COMPLETE ARMS. BARRELS and LOCKS (Nov. 1699 only).

R

RABY, Alexander. London. ID Mar. 1794 – FD May 1803.
  MERCHANT who supplied large quantities of non-regulation (?) military-pattern arms purchased in the open market. From Feb. 1797 ROUGH STOCKED & SET UP lesser numbers of regulation pattern arms.

RADBIRD, James. London. Dec. 1703 only.
  COMPLETE ARMS on account of Andrew Dolep.

REA, John. London. ID July 1793 – FD May 1808.
  LOCKS, BARRELS, ROUGH STOCKED & SET UP, and COMPLETE ARMS.
  Also purchased non-regulation military-pattern arms in the open market.

REDDELL, Joseph H. Birmingham. ID May 1805 – May 1809; July 1815 – July 1816; Nov. – FD Dec. 1823. BAYONETS, chiefly rifle sword-bayonets. In and out of partnership with Thomas Bate before and after the above dates. BARRELS and LOCKS, Pattern 1823 Infantry Rifle only in 1823.

REDDELL & BATE. Birmingham. ID Aug. 1804 – FD June 1807.
  BAYONETS, SWORDS. Partnership dissolved and the two partners delivered separately.

REYNOLDS, Thomas. London. ID Oct. 1795 – Aug. 1809; Mar. 1825; May 1827;– FD 1836.
  LOCKS, BARRELS, ROUGH STOCKED & SET UP and COMPLETE ARMS. Also purchased non-regulation military-pattern arms in the open market. Pattern 1823 Infantry Rifles in 1825-7; 16" Cavalry Carbines for the Queen of Spain in 1835-6; infantry rifles for the Shah of Persia in 1836, also carbines and pistols in 1836.

REYNOLDS, William. London. ID June 1827; FD Feb. 1828.
  SMALLWORK. Supplied Heavy Dragoon Carbine ribs and general smallwork for Light Infantry Serjeant's Carbines to RSAF Enfield.

REYNOLDS & SON, Thomas. London. ID Aug. 1836 – FD July 1838.
  ROUGH STOCKED & SET UP. One delivery of Pattern 1823 Infantry Rifles (Shah of Persia) in 1836 and one of pistols and carbines (1837) and a final delivery of infantry rifles in 1838.

ROCK, John. Birmingham. ID Feb. 1804 – FD Dec. 1816.
  RAMRODS, BAYONETS.

ROCK, Joseph & James. Birmingham. ID Feb. 1816; FD Dec. 1816.
  LOCKS, BARRELS, BAYONETS.

ROCK, Joseph & Samuel. ID Feb. 1816 – FD 1824.
  BAYONETS. Take over their late mother Martha's contract. BARRELS, Pattern 1823 Infantry Rifle only in 1823-4.

ROCK, Martha. Birmingham, ID Nov. 1810 – FD Jan. 1816.
  BARRELS, LOCKS, RAMMERS and BAYONETS. Takes over contracts from her late husband William.

ROCK, Samuel. Birmingham. Nov. 1823 only.
   LOCKS, Pattern 1823 Infantry Rifle.

ROCK, William. Birmingham. ID June 1799 – FD Nov. 1810.
   BARRELS, LOCKS, RAMMERS and BAYONETS.

ROLFE, Elizabeth. Birmingham. ID Feb. 1815 – FD Dec. 1816
   ROUGH STOCKED & SET UP, COMPLETE ARMS, including 1813 Indian arms. Widow who took over late husband's contracts.

ROLFE, William Isaac. Birmingham. ID Aug. 1810 – FD Jan. 1815.
   ROUGH STOCKED & SET UP, COMPLETE ARMS, including 1813 Indian arms.

ROSE, William. London. ID Jan. 1706 – FD Aug. 1720.
   COMPLETE ARMS to May 1711. STOCKER & SETTER UP from Mar. – Sept. 1715; one delivery of LOCKS in 1720.

ROSS, Robert. London. ID Sept. 1778 – FD Apr. 1779.
   ROUGH STOCKER and SETTER UP of two batches of Short Land muskets.

ROUND, Benjamin. Birmingham. ID July 1812 – Feb. 1818; Apr. 1823 – FD 1824.
   BARRELS. Pattern 1823 Infantry Rifle only in 1823–4.

ROUND, Joseph. Birmingham. ID Sept. 1801 – Feb. 1818; FD Dec. 1823.
   LOCKS. Pattern 1823 Infantry Rifle only in 1823.

RUBERY, J. Birmingham. Apr. 1838 only.
   LOCKS, Pattern 1823 Infantry Rifle.

RUSSELL, James. Birmingham. ID Apr. 1804 – FD 1815.
   BARRELS.

RUSSELL, John. Birmingham. ID Jan. 1794 – Feb. 1818; FD Dec. 1823.
   BARRELS. Son of Thomas, in partnership with him from Nov. 1813 but continued to be billed in his own name. Pattern 1823 Infantry Rifle only in 1823.

RUSSELL, Thomas. Birmingham. ID Apr. 1793 – FD 1815.
   BARRELS. Took son John into partnership, Dec. 1809.

S

SALE, Edward. London. ID Apr. 1715; FD Aug. 1716.
   STOCKER & SETTER UP.

SALE, Edward. London. ID May 1743 – FD June 1749.
   SETTER UP. One of eight new setters up taken on beginning in 1742 to expand wartime production.

SALTER, George. Birmingham. ID July 1812 – FD Feb. 1818.
   BAYONETS.

SALTER, John. Birmingham. ID Oct. 1808; FD Feb. 1818.
   BAYONETS (1809), BARRELS (1816).

SANSOM, S. Birmingham. July 1838 only.
   LOCKS, Pattern 1823 Infantry Rifle.

SARGANT & SON. Birmingham. ID 1836; FD Dec. 1837.
ROUGH STOCKED & SET UP, pistols and carbines.

SAUNDERS, Thomas. London. ID Oct. 1688 – FD Mar. 1716.
COMPLETE ARMS to June 1711. BARRELS and LOCKS in Nov. 1699 and LOCKS for Irish service muskets in Dec. 1713. STOCKER & SETTER UP from Dec. 1714.

SAVAGE, Richard. London. ID Oct. 1688 – FD Sept. 1692.
COMPLETE ARMS.

SHARP, Richard. London. ID May 1743 – FD pre-Aug. 1755.
Gun ENGRAVER to the Board of Ordnance. Succeeded by his brother William.

SHARP, William. London. ID June 1750 – FD died Feb. 1786.
Gun ENGRAVER to the Board of Ordnance. Also supplied SMALLWORK, ROUGH FORGINGS of LOCK PARTS, TAILPIPE SPRINGS, and steel RAMMERS. Also small deliveries of BARRELS in 1759 and 1762. Also all sorts of specialist gunsmithing workshop tools, and stamps for marking metal parts. Took son William as partner, Sharp & Son, in Dec. 1781.

SHARP, William (II). London. ID Dec. 1781; FD not established.
Gun ENGRAVER to the Board of Ordnance. Worked with his father William (I) until his death in Feb. 1786, then on his own (?).

SHARP, William & John. London. ID Dec. 1783; FD 1786.
Gun ENGRAVERS to the Board of Ordnance.

SHENSTONE, Hannah. Birmingham. Sept. 1823 only.
LOCKS, Pattern 1823 Infantry Rifle.

SHENTON, William. Birmingham. ID Mar. 1804 – FD Feb. 1818.
LOCKS.

SHERWOOD, Elizabeth & William. London. May 1815 only.
LOCKS, New Land Pattern musket.

SHERWOOD, James. London. Jan. 1808 only.
LOCKS, New Land Pattern musket.

SHERWOOD, John junior & James. London. ID July 1807 – FD Apr. 1810.
LOCKS, New Land Pattern musket.

SHERWOOD, Joseph senior. London. ID Dec. 1796 – FD May 1815.
COMPLETE ARMS, sometimes in partnership with Walter Dick, until June 1799. Also LOCKS, New Land Pattern musket from Jan. 1803; given a contract to deliver 150 locks per month in Sept. 1806. Rifle box tools from Feb. 1801.

SHERWOOD, Joseph junior & James. London. ID June 1805 – FD June 1806.
LOCKS, New Land Pattern musket.

SHERWOOD, J. & W. London. ID Sept. 1823 – July 1826; FD Apr. 1828.
LOCKS, Pattern 1823 Infantry Rifle. In 1828 supplied locks for Pattern 1798 Heavy Dragoon Carbines and Pistols to RSAF Enfield.

SHOREY, Joseph. London. ID June 1702 – FD Mar. 1711.
COMPLETE ARMS.

SIBLEY, John. London. ID Oct. 1688 – FD June 1711.
  COMPLETE ARMS including Government-gift New York Indian fusils in Oct. 1702.

SIBTHORPE, Robert. London. ID June 1706 – FD May 1715.
  COMPLETE ARMS to June 1712 including Government-gift New York Indian fusils in Feb. 1710. A musket and LOCKS for Irish service in Dec. 1713. STOCKER & SETTER UP May 1715 only.

SIDDON, William. Birmingham. ID Sept. 1801 – FD July 1813.
  LOCKS.

SIDDON & CO. Birmingham. Dec. 1823 only.
  LOCKS, Pattern 1823 Infantry Rifle.

SIDDON & SONS. Birmingham. ID Aug. 1813 – FD Feb. 1818.
  LOCKS.

SILKE, John. London. ID Apr. 1689 – FD Sept. 1705.
  COMPLETE ARMS, and BARRELS (Nov. 1699 only).

SILKE, Robert (1). London. ID Oct. 1688 – FD June 1700.
  COMPLETE ARMS, many of them Sea Service, including Government-gift New York Indian fusils in Aug. 1699 and June 1700. BARRELS and LOCKS (Nov. 1699 only). Did much work in teams with other makers, particularly Thomas Austin and George Fisher. In Feb. 1689 organized a group of gunmakers to go to Ireland to work.

SILKE, Robert (2). London. ID May 1694 – FD Apr. 1695.
  COMPLETE ARMS.

SIMES, John. Birmingham. ID Apr. 1800 – FD July 1805.
  BARRELS.

SIMES & PRIEST. Birmingham. ID Mar. 1804; FD 1811.
  BARRELS.

SIMPSON, William. London. ID Feb. 1718 – FD June 1720.
  STOCKER & SETTER UP of Pattern 1718 muskets.

SINCKLER, Richard. London. ID Mar. 1715 – Jan. 1720; Feb. 1723; FD Apr. 1728.
  STOCKER & SETTER UP from Mar. 1715, including Government-gift New York Indian fusils in July 1716, some of which he also cleaned and repaired after their loss by shipwreck. COMPLETE regimental arms 1727-8. In July 1727 he supplied muskets and bayonets to Brigadier Kirk's Regiment; in April 1728 supplied muskets and bayonets and pairs of pistols to Major Honywood's 11th Dragoons.

SLATER, John. Birmingham. ID Oct. 1823 – Aug. 1826; FD Feb. 1828.
  LOCKS, Pattern 1823 Infantry Rifle. In 1828 supplied locks for Pattern 1798 Heavy Dragoon Carbines and Pistols to RSAF Enfield.

SMART, Francis. London. ID Apr. 1695 – FD Mar. 1721.
  COMPLETE ARMS to Sept. 1712 including Government-gift New York Indian fusils in Feb. 1710. BARRELS and LOCKS (Nov. 1699). LOCKS (1708-9) STOCKER & SETTER UP from Feb. 1715.

SMART, John. London. ID Mar. 1720 – FD Aug. 1720.
  STOCKER & SETTER UP of Pattern 1718 muskets.

SMITH, James. Birmingham. ID Nov. 1793 – Feb. 1818; Sept. 1823 – Sept.1826; FD Feb. 1828.   LOCKS. Pattern 1823 Infantry Rifle only 1823-6.   In 1828 supplied locks for Pattern 1798 Heavy Dragoon Carbines and Pistols to RSAF Enfield.

SMITH, John. Birmingham. ID Aug. 1736 – FD Sept. 1748.
LOCKS and LOCK FORGINGS.

SMITH, William. London. Mar. 1708 only.
COMPLETE ARMS,   one delivery of Pattern 1703 Muskets.

SMITHETT, Dorothy. London. ID Mar. 1718 – FD June 1720.
STOCKER & SETTER UP of Pattern 1718 muskets.

SMITHETT, George senior. London. ID Dec. 1703 – FD Dec. 1716.
COMPLETE ARMS to June 1711. STOCKER & SETTER UP from Mar. 1715, including Government-gift New York Indian fusils in July 1716.

SMITHETT, George senior & junior. London. ID July 1717 – FD Mar. 1720.
STOCKER & SETTER UP, chiefly of Pattern 1718 muskets.

SMITHETT, George junior. London. June 1716 only.
STOCKER & SETTER UP, one delivery.

SOWERBY, Joseph. London. ID Dec. 1716 – FD Mar. 1720.
STOCKER & SETTER UP.

SOWERBY, William. London. ID Mar. 1708 – June 1720;   FD June 1728.
COMPLETE ARMS to Sept. 1712 including Government-gift New York Indian fusils in Feb. 1710 and July 1716.   Muskets, BARRELS and LOCKS for Irish service in Mar. 1714.   STOCKER & SETTER UP from Dec. 1714.   Important supplier.   One delivery in 1728.

SPENCER, Elizabeth. London.   Nov. 1699 only.
BARRELS and LOCKS.

SPENCER, James. London. ID June 1689 – FD Mar. 1696.
COMPLETE ARMS.

SPITTLE, James. Birmingham. Dec. 1812 only.
LOCKS.

SPITTLE, Joseph. Birmingham. ID Mar. 1806 – FD Feb. 1818.
LOCKS.

SPITTLE, M. & F. Birmingham. ID Sept. 1823; FD Feb. 1824.
LOCKS,   Pattern 1823 Infantry Rifle only.

SPITTLE, Peter. Birmingham. ID Jan. 1801 – Feb. 1818;   FD Oct. 1823.
LOCKS.   Pattern 1823 Infantry Rifle only in 1823.

SPITTLE & NEGUS. Birmingham.   Mar. 1804 only.
LOCKS. Signed the open agreement of contractors with the Board, but no record of deliveries under this style.

STACE, Joseph. London. ID Oct. 1688 – FD Apr. 1689.
COMPLETE ARMS.

STAMPS, Thomas. Birmingham. Dec. 1767 only.
LOCKS, Land Pattern musket, one delivery.

STEBBEN, Robert. London. ID Apr. 1695 – FD Nov. 1710.
   COMPLETE ARMS. BARRELS and LOCKS (Nov. 1699), LOCKS 1703.

STEVENS, Andrew. London. ID June 1708 – FD Dec. 1711.
   COMPLETE ARMS and LOCKS (1708-9, 1711).

STOKES, Thomas. Birmingham. ID May 1798 – Feb. 1818; FD Oct. 1823.
   BARRELS. Struck off the list of suppliers Jan. 1809 for not having supplied for over a year, but was subsequently re-admitted in Mar. 1812. Pattern 1823 Infantry Rifle only in 1823.

STONE, Guy. Birmingham. ID Jan. 1708 – FD June 1711.
   SOCKET BAYONETS.

STONE, Thomas. Birmingham. ID Mar 1806 – Feb 1818; Sept 1823 – Aug.1826; FD Mar 1828.
   LOCKS. Also supplying gunlock forgings to Dublin and Lewisham in Jan. 1811. Pattern 1823 Infantry Rifle only 1823-6. In 1828 supplied locks for Pattern 1798 Heavy Dragoon Carbines and Pistols to RSAF Enfield.

STRINGER, Elizabeth. London. ID Mar. 1715 – FD Sept. 1715.
   STOCKER & SETTER UP. Widow of Ralph.

STRINGER, Ralph. London. ID Apr. 1695 – FD Oct. 1711.
   COMPLETE ARMS. BARRELS and LOCKS (Nov. 1699 only).

SUTHERLAND, Ramsay & Richard. Birmingham. ID Feb. 1809 – May 1818; Oct. 1823; FD July 1825. BARRELS, LOCKS, ROUGH STOCKED & SET UP, COMPLETE ARMS including 1813 Indian arms. Third largest producer of India Pattern muskets, after the two Ketland firms. Produced all of the Pattern 1810 musket-bore Infantry Rifles. Pattern 1823 Infantry Rifle locks only in 1823, Pattern 1823 Infantry Rifles in 1825.

T

TAYLOR, George. London. ID Dec. 1701 – FD Sept. 1711.
   COMPLETE ARMS including Government-gift New York Indian fusils in Feb. 1710. A bill of 8 Jan. 1709 was paid to Mary, and one of 2 Dec. 1710 was in both names.

TAYLOR, Godfrey. London. ID Oct. 1688 - FD June 1700.
   COMPLETE ARMS including Government-gift New York Indian fusils in June 1700. LOCKS in Nov. 1699 only.

TAYLOR, John. London. ID June 1747 – FD June 1749.
   SETTER UP. The latest of eight firms added from 1742 to increase wartime output.

TAYLOR, Thomas. London. ID July 1710 – FD Nov. 1710.
   COMPLETE ARMS, Pattern 1703 Muskets.

TAYLOR & DAVIES (DAVIS). Birmingham. ID May 1799 – FD Feb. 1804.
   BARRELS. Became William Davies. Also two deliveries in 1805 by Martha Taylor & Son.

THOMAS, Ralph. London. ID Mar. 1696 – FD June 1702.
   COMPLETE ARMS. A bill of Sept. 1703 was in the name of Ralph and Elizabeth, one of Jan. 1705 to Elizabeth per William Collins, and another of the same date to Elizabeth Thomas.

THOMAS, William. London. ID Sept. 1838; FD Feb. 1839.
  ROUGH STOCKED & SET UP,   Pattern 1823 Infantry Rifles.

THOMPSON (THOMSON) James. London. ID July 1802 – FD Mar. 1808.
  ROUGH STOCKED & SET UP. Became '& Son', Apr. 1808.

THOMPSON, John. London. ID Jan. 1716 – FD Dec. 1728.
  Wooden RAMRODS.

THOMPSON & SON, James. London. ID Apr. 1808 – Sept. 1819; Jan. 1821; Feb. 1824; Dec. 1825; May 1827; 1835 – FD Aug. 1836.   ROUGH STOCKED & SET UP. Infantry rifles only 1821-36, the 1836 delivery for the Shah of Persia.   16" Cavalry Carbines for the Queen of Spain in 1835–6.

THORNHILL, Charles. Birmingham. ID Oct. 1823 – Sept. 1826; FD Mar. 1828.
  LOCKS, Pattern 1823 Infantry Rifle. In 1828 supplied locks for Pattern 1798 Heavy Dragoon Carbines and Pistols to RSAF Enfield.

TIPPIN & EDGE. Birmingham. ID Sept. 1747; FD Sept. 1748.
  LOCKS, BARRELS and one delivery of COMPLETE Pattern 1742 Long Land Muskets and bayonets (1747).   Partnership of Walter Tippin and Richard Edge.

TITTENSOR, John. Birmingham. ID Nov. 1707 – FD Mar. 1718.
  COMPLETE ARMS to Jan. 1711.   BARRELS (1707-8).   Barrel repair work Mar. 1711-June 1716.   From Mar. 1715 also STOCKER & SETTER UP.

TITTENSOR, Joseph. Birmingham. Sept. 1710 only.
  COMPLETE ARMS, one delivery of Pattern 1703 muskets.

TITTENSOR, William. Birmingham. ID Mar. 1708 – FD Mar. 1711.
  BARRELS (1708-9) and two deliveries of COMPLETE ARMS 1710-11.

TOMKYS & SHORT. Birmingham. ID Sept. 1757 – FD Dec. 1768.
  BARRELS and LOCKS, originally as executors of John Willets.

TOUGH, Mary. London. ID Dec. 1694 – FD Sept. 1718.
  COMPLETE ARMS to Mar. 1711 including Government-gift New York Indian fusils in Feb. 1710.   BARRELS and LOCKS (Nov. 1699 only).   Muskets and barrels for Irish service in Dec. 1713.   STOCKER & SETTER UP from Dec. 1714, important supplier.

TOUGH, Robert. London. ID Mar. 1705 – FD Aug. 1721.
  COMPLETE ARMS to Mar. 1711. STOCKER & SETTER UP from Mar. 1715   including Government-gift New York Indian fusils in July 1716.

TOWLE, Elizabeth. London. ID Feb. 1693 – FD Mar. 1716.
  COMPLETE ARMS to Jan. 1712 including Government-gift New York Indian fusils in Feb. 1710.   Muskets, barrels and locks for Irish service Dec. 1713.   STOCKER & SETTER UP from Dec. 1714. Being paid for arms from Thomas beginning 28 Feb. 1691.

TOWLE, Thomas (I). London. ID Apr. 1688 – FD Feb. 1691.
  COMPLETE ARMS.

TOWLE, Thomas (II). London. ID Dec. 1703 – FD June 1720.
  COMPLETE ARMS to May 1711 including Government-gift New York Indian fusils in Feb. 1710. STOCKER & SETTER UP from Dec. 1714.

TRANTER, William. Birmingham. ID July 1812 – FD Feb. 1818.
    RAMRODS.

TRESTED, Richard. London. ID Nov. 1777 – FD May 1786.
    ROUGH STOCKER.

TREVEY, Nathaniel. London. ID July 1743 – FD June 1749.
    SETTER UP. One of eight new contractors taken on beginning in 1742 to increase wartime production.

TRULOCK (TRUELOCK), George. London. ID Oct. 1695 – FD Oct. 1704.
    COMPLETE ARMS. A bill of Feb. 1705 paid per Gertrude. In July 1697, Trulock produced *'Five double Fuzees by him & delivered into his Majties Stores for the use of my Lord Bellomont to carry with him to his Government'* [as Governor of the Province of New York].

TRUMAN (TRUEMAN), Elizabeth. Birmingham. ID June 1801–Feb. 1818; Jan.1823–FD 1824.
    BARRELS. Pattern 1823 Infantry Rifle only 1823-4. Widow of John.

TRUMAN (TRUEMAN), John. Birmingham. ID May 1798 – FD May 1801.
    BARRELS.

TUCKER, Thomas. London. ID Jan. 1777 – Feb. 1783; FD Oct. 1785.
    ROUGH STOCKER. Also COMPLETE Sea Service muskets Mar. – May 1780. One small delivery of rough stocked carbines in 1785.

TURNER, B. Birmingham. Mar. 1838 only.
    LOCKS, Pattern 1823 Infantry Rifle.

TURNER, John. London. Apr. 1695 only.
    COMPLETE ARMS.

TURNER, Joseph. Birmingham. ID Nov. 1823; FD Oct. 1838.
    BARRELS, Pattern 1823 Infantry Rifle.

TURTON, Joseph. Birmingham. Aug. 1825 only.
    BARRELS, Pattern 1823 Infantry Rifle.

TURVEY, Edward. London. ID July 1690 – FD Mar 1716.
    COMPLETE ARMS to June 1712 including Government-gift New York Indian fusils in Feb. 1710. BARRELS and LOCKS in Nov. 1699. LOCKS (1708). Muskets for Irish service in Dec. 1713. STOCKER & SETTER UP from Feb. 1715.

TURVEY, John. London. ID July 1690 – FD Dec. 1701.
    COMPLETE ARMS. BARRELS and LOCKS (Nov. 1699 only). Edward was John's executor in Dec. 1703.

TURVEY, William. London. Feb. 1721 only.
    SETTER UP of one delivery of Pattern 1718 muskets.

TYLOR (TYLER), Thomas. London. ID Oct. 1688 – FD Dec. 1715.
    COMPLETE ARMS to June 1711. STOCKER & SETTER UP Dec. 1715.

V

VAUGHAN, John. London. ID Sept. 1704 – June 1720; May 1728 – FD Sept. 1747.
COMPLETE ARMS to June 1712. LOCKS (1704-9) and for Irish service muskets Dec. 1713. ROUGH STOCKER and SETTER UP from Dec. 1714. A major supplier of LOCKS May 1728 – Sept. 1747. Died 1747.

VAUGHTON, Humphrey. Birmingham. Mar. 1710 only.
BARRELS. One delivery of musket and carbine barrels.

VAUGHTON, Ryland. Birmingham. Dec. 1709 only.
BARRELS. One delivery of musket and carbine barrels.

VAUGHTON, Samuel. Birmingham. ID Mar. 1709 – FD Dec. 1711.
COMPLETE ARMS and BARRELS.

VERNON, George. Birmingham. ID June 1757 – FD Dec. 1758.
LOCKS. Subsequently in partnership with George Haskins, as Haskins & Vernon.

W

WAGG, Francis. London. ID Oct. 1688 – FD Mar. 1696.
COMPLETE ARMS. Bills of July 1690, Feb. 1691 and Jan. 1693 were paid to Margaret, with subsequent dates to Francis, suggesting that there may have been Francis (I) and Francis (II).

WALLER, James. London. ID May 1767 – FD May 1781.
ROUGH STOCKER. In partnership with father Richard Feb.1755 – Sept. 1767, took over business 1767. Divided rough stocking with Joseph Loder from March 1769 until 1777 when other contractors were added.

WALLER, Richard. London. ID Dec. 1718 – FD Sept. 1767.
GUN STOCK MAKER, ROUGH STOCKER, WOOD SUPPLIER. In partnership with son James 1755-67. Had monopoly of rough stocking for the Board from ca. 1730 to 1756, and virtual monopoly until his death in Feb. 1769.

WARD, Richard. London. ID Aug. 1718 – FD Sept. 1719.
ROUGH STOCKER of Pattern 1718 arms.

WARD, Thomas. London. ID Apr. 1695 – FD Dec. 1706.
COMPLETE ARMS, BARRELS and LOCKS (Nov. 1699) and LOCKS (1704).

WARNER & SON. London. ID Mar. 1797 – FD Sep. 1801.
BRASS GUN FURNITURE.

WARREN, Charles. London. Mar. 1696 only.
COMPLETE ARMS. Delivered five musketoons.

WATKINSON, John. London. ID Apr. 1688 – FD Jan. 1693.
COMPLETE ARMS. Widow Mary took over contracts.

WATKINSON, Mary. London. ID Apr. 1695 – FD Jan. 1705.
    COMPLETE ARMS. Bills dated Sept. 1692 and Nov. 1693 were paid to Mary for John. BARRELS and LOCKS (Nov. 1699 only).

WELFORD, Richard. London. ID Feb. 1715 – FD Dec. 1720.
    STOCKER & SETTER UP, of Pattern 1718 muskets from June 1719.

WESTON, Edmond (Edmund). London. ID July 1690 – FD Sept. 1711.
    COMPLETE ARMS. BARRELS and LOCKS (Nov. 1699 only).

WESTON, Edward. London. 1717 only
    IRON GUN FURNITURE, one delivery of 300 sets.

WESTON, Richard. London. ID Mar. 1700 – FD Dec. 1710.
    COMPLETE ARMS, BARRELS and LOCKS (1699).

WHATELY, Henry P. Birmingham. ID May 1804 – FD 1815.
    BARRELS.

WHATELY, Henry & John. Birmingham, ID Oct. 1803 – FD Dec. 1816.
    BARRELS, LOCKS, RAMMERS, ROUGH STOCKED & SET UP and COMPLETE ARMS including 1813 Indian arms.

WHATELY, John junior. Birmingham. ID Jan. 1757 – FD Oct. 1757.
    BARRELS.

WHATELY, John (younger). Birmingham. ID Dec. 1776 – Jan. 1788; Feb. 1794 – FD Mar. 1804.
    BARRELS, LOCKS and RAMMERS. Also COMPLETE ARMS from 1794.

WHATELY, John. Birmingham. ID Jan. 1817 – Sept. 1819; FD Dec. 1823.
    ROUGH STOCKED & SET UP, LOCKS and BARRELS. Successor to Henry & John. Pattern 1823 Infantry Rifle barrels only in 1823.

WHATELY & SON, John (elder & younger). Birmingham. ID Nov. 1757 – FD 1776.
    BARRELS, LOCKS (1762-4) and RAMMERS. One of the principal suppliers during this period.

WHEELER, John. Birmingham. Dec. 1823 only.
    BARRELS, Pattern 1823 Infantry Rifle.

WHEELER, Robert. Birmingham. Dec. 1823 only.
    BARRELS, Pattern 1823 Infantry Rifle.

WHEELER, Robert junior. Birmingham. ID May 1797 – FD June 1808.
    BARRELS (from Sept. 1800), LOCKS (to June 1800), BAYONETS, ROUGH STOCKED & SET UP and COMPLETE ARMS. Became Wheeler & Son.

WHEELER & SON, Robert. Birmingham. ID July 1808 – Sept. 1819; Oct. 1823; July 1825; Jan. 1826; 1835 – FD 1837. BARRELS, LOCKS, RAMMERS, BAYONETS ROUGH STOCKED & SET UP and COMPLETE ARMS including 1813 Indian arms. Senegal trade muskets in 1815. Pattern 1823 Infantry Rifle locks only in 1823 and 1826, Pattern 1823 Infantry Rifles only in 1825; pistols and carbines 1835-7.

WHITE, John. London. ID June 1716 – FD Dec. 1720.
    STOCKER & SETTER UP including Government-gift New York Indian fusils in July 1716, but chiefly of Pattern 1718 arms.

WHITE, Thomas. London. ID June 1702 – FD Oct. 1716.
COMPLETE ARMS to May 1711. BARRELS and LOCKS for Irish service muskets in Dec. 1713. STOCKER & SETTER UP from Mar. 1715.

WHITEHEAD, Francis. London. ID Dec. 1801 – FD 1808.
LOCKS.

WHITEHEAD, J. London. Mar. 1838 only.
LOCKS, Pattern 1823 Infantry Rifle.

WHITEHEAD & CO. London. ID Sept. 1823; May 1825; Oct. 1826; FD Feb.1828.
LOCKS, Pattern 1823 Infantry Rifle. In 1828 supplied locks for Pattern 1798 Heavy Dragoon Carbines and Pistols to RSAF Enfield.

WHITEHEAD & SON, Francis. London. ID Aug. 1806 – FD Feb. 1818.
LOCKS.

WILKES (WILKS), Job. Birmingham. ID Mar. 1806 – Feb. 1818; Sept. 1823; June 1825; FD Aug. 1826. LOCKS. Pattern 1823 Infantry Rifle only from 1823.

WILKES, James. London. ID May 1799 – FD 1800.
LOCKS, small number of COMPLETE ARMS, including a breech-loading rifle which was rejected by the Board in 1805.

WILLETTS (WILLETS, WILLET). Benjamin. Birmingham. ID June 1769 – Dec. 1782; Mar. 1794 – FD Mar. 1795. BARRELS, LOCKS, and COMPLETE Pattern 1776 muzzle-loading, and Ferguson Rifles. Executors completed contracts by Dec. 1796.

WILLET(S), John. Wednesbury. ID Sept. 1743 – Mar. 1748; Aug. 1757 – FD Dec. 1758.
BARRELS and LOCKS. Executors were Tomkys & Short.

WILLETTS, Mary. Birmingham. ID Aug. 1797 – FD Mar. 1804.
BARRELS, LOCKS, COMPLETE ARMS. Supplied pattern muskets, locks and rammers for the New Land Pattern muskets, Aug. 1802.

WILLETTS & HOLDEN. Birmingham. ID Jan. 1797 – Oct. 1797; Feb. 1805 –.Sept. 1819; Mar., Sept. 1823 – FD 1824. BARRELS, LOCKS, ROUGH STOCKED & SET UP (from 1805), and COMPLETE ARMS including 1813 Indian arms. The Sept. 1823 delivery is Pattern 1823 Infantry Rifle locks only, the Sept. 1823–4 deliveries are for Pattern 1823 Infantry Rifle barrels only, all in the name of Willetts & Co.

WILLIAMS, John. London. ID Dec. 1701 – FD June 1720.
COMPLETE ARMS to June 1711. Muskets and locks for Irish service in Dec. 1713. LOCKS (1715-20). STOCKER & SETTER UP from Mar. 1715.

WILLMORE, Joseph. Birmingham. ID Apr. 1813 – FD Feb. 1818.
RAMRODS.

WILLOWES, John. London. ID Mar. 1705; Feb. 1714 – FD Mar. 1720.
COMPLETE ARMS, one delivery in 1705. STOCKER & SETTER UP from June 1715 including Government-gift New York Indian fusils in July 1716.

WILSON, C. Birmingham. May 1838 only.
LOCKS, Pattern 1823 Infantry Rifle.

WILSON, Richard. London. ID June 1746 – Oct. 1749; June 1756 – FD July 1761.
SETTER UP, also supplied BARRELS and LOCKS from 1756.

WILSON, William (I). London. ID Jan. 1777 – FD Jan. 1782.
SETTER UP. Son of Richard, in partnership with father from 1755 until Richard's death in 1766. No Ordnance work until the above date.

WILSON & SON, William. London. ID Oct. 1803 – FD Aug. 1805.
ROUGH STOCKED & SET UP India Pattern muskets. Partnership of William (I) and his son William (II).

WINESOP (WINESUP), John. London. ID Mar. 1696 – Dec. 1706; Feb. 1715 – FD June 1715.
COMPLETE ARMS to Dec. 1706. BARRELS and LOCKS (Nov. 1699 only), LOCKS (1706). STOCKER & SETTER UP from Feb. 1715.

WITTON, Joseph. London. ID May 1786 – FD 1806.
Gun ENGRAVER to the Board of Ordnance, the above initial date being that of his appointment.

WOOD, John. London. ID Mar.– Sept. 1748; Dec. 1756; FD Feb. 1757.
LOCKS, Pattern 1740 musket only. Two deliveries of Old Pattern musket (i.e. 1740) in 1756-7.

WOOD, Joseph. Birmingham. ID Sept. 1800 – FD 1801.
LOCKS, almost entirely Pattern 1800 Infantry Rifle.

WOODCRAFT, John. Birmingham. ID Dec. 1703 – FD Aug. 1711.
SOCKET BAYONETS.

WOODELL, George. London. Aug. 1719 only.
ROUGH STOCKER, one delivery of Pattern 1718 Sea Service muskets.

WOOLDRIDGE, Richard. London. ID 1703 – FD 1745.
COMPLETE ARMS including PATTERN ARMS, BARRELS, LOCKS, BULLET MOULDS. Master Furbisher of the Small Gun Office establishment in the Tower from June 1707. A working gunsmith in his own right he fabricated '*a fine Gun for the King*' in 1725 with silver mounts and wirework, and with a gold touch-hole. He made, or superintended the making of, a variety of arms for regimental colonels, his last order of this nature being in Nov. 1743 when he produced 66 pairs of pistols for Gen. Churchill's Dragoons. Died 1745 and replaced as Master Furbisher by Thomas Hatcher. His son Richard junior was working in the Small Gun Office during the latter part of 1722, and may have signed arms as well as his father. His work needs to be distinguished from that of his father.

WOOLEY, James. Birmingham. ID June 1799 – FD Feb. 1818.
BAYONETS including sword bayonets for infantry rifles, and RAMRODS. In June 1813 supplied India Pattern musket bayonets with steel necks.

NOTE: the precise identification and differentiation between the various firms in which James Wooley participated is impossible to achieve on the basis of Ordnance records alone. The chief reason is that the clerks often shortened lengthy firm styles to '& Co.' or even '&c' because they knew to whom they referred: we do not. Spelling of the surname varies without obvious significance between Wooley and Woolley.

WOOLEY, DEAKIN, DUTTON & JOHNSON. Birmingham. ID Feb. 1805; FD ?
BAYONETS, sword bayonets.

WOOLLEY, DEAKIN & DUTTON. Birmingham. ID Oct. 1809; FD (?)
BAYONETS.

WOOLLEY & CO. Birmingham. ID June 1799; ID May 1828.
BARRELS, LOCKS, BAYONETS, COMPLETE ARMS. In 1828 supplied RAMMERS for Heavy Dragoon Carbines and Pistols, and Infantry Rifles.

WOOLLEY & DEAKIN. Birmingham. ID Apr. 1803; FD ?
BAYONETS.

WOOLLEY & PRICE. Birmingham. Nov. 1823 only.
LOCKS, Pattern 1823 Infantry Rifle. This is almost certainly the same firm listed under its earlier style of Woolley, Price & Jones.

WOOLLEY & SARGEANT. Birmingham. ID Aug., Oct., FD Nov. 1823.
BARRELS, Pattern 1823 Infantry Rifle in Aug. and Nov. LOCKS, Pattern 1823 Infantry Rifle in Oct.

WOOLLEY, PRICE & JONES. Birmingham. ID Mar. 1809 – FD Oct. 1816.
LOCKS. Partnership which set up a barrel-making and cock-stamping factory which began supplying the Ordnance. First order for pressed cocks for (New) Land muskets Dec. 1811 through Apr. 1816, supplied at least 66,200.

WRIGHT, Robert. London. ID June 1797 – June 1802; Nov. 1820; Feb. 1824; Feb. 1825; FD May 1827. ROUGH STOCKED & SET UP and COMPLETE ARMS. All 1820s deliveries were Infantry Rifles.

WRIGHT, Thomas. London. ID Oct. 1688 – Dec. 1706; FD Mar. 1715.
COMPLETE ARMS to Dec. 1706. BARRELS and LOCKS (Nov. 1699 only). STOCKER & SETTER UP Mar. 1715.

WRIGHT & CO., Robert. London. ID May 1801 – FD Sept. 1819.
ROUGH STOCKED & SET UP and COMPLETE ARMS.

Y

YATES, J. Birmingham. Feb. 1838 only.
LOCKS, Pattern 1823 Infantry Rifle.

YEOMANS, James. London. ID Aug. 1809; FD Aug. 1836
COMPLETE ARMS, Extra Service muskets, one delivery. Previously and subsequently employed for cleaning and repairing arms. ROUGH STOCKED & SET UP Pattern 1823 Infantry Rifles for the Shah of Persia in 1836.

YEOMANS & SON, James. London. ID 1835 – FD Feb. 1839.
ROUGH STOCKED & SET UP 16" Cavalry Carbines for the Queen of Spain in 1835, pistols and carbines in 1836-7, Pattern 1823 Infantry Rifles 1838-9.

PART 2

BOARD of ORDNANCE SMALL ARMS COMPONENTS & PRODUCTION CONTRACTORS

Listed by the time periods during which they worked, and within each time-span by product, 1688 – 1840

Within each time frame the contractor is listed under the component or service which they performed during this time frame only, and the dates there given are for that service, not for work they may have done previously or subsequently, or in another product or service category during the same time period.

The dates given are in each case the date of the bill for completed work. A ✷ indicates that a contractor worked in the preceding and/or succeeding time-span.

1688 – 1715

During this period it was the general practice for the contractors, largely members of the Worshipful Company of Gunmakers of London, to supply complete arms. The number of arms required in a given category (musket, carbine, pistol) would be settled by the Board, and a pattern and price agreed by negotiation between the Board and the Gunmakers' Company. The officials of the Company would then distribute the work to be done amongst their members according to the capacity of the individual maker and his standing with the officials of the Company. There was only one large components order during this period, in November 1699, as a result of which some 4,250 46-inch musket barrels and some 3,700 musket locks were delivered by the contractors into Ordnance Stores.

The Irish material contracted for in August 1713 represents the earliest recorded instance of English contractors supplying components intended for use by gunsmiths in the Dublin Establishment. It was a very small order, perhaps intended as a pricing guide.

On 15 September 1714, the first order for arms under the new 'Ordnance System' was awarded. This was a Contract for His Majesties Land Service musquet barrels to be stocked and set up with His Majesties locks and brasswork according to the pattern, at 8/9 each. This meant that the barrels, locks and brass furniture would be issued to the contractors from Ordnance Stores and assembled into complete arms in two stages, firstly rough stocking, and secondly setting up, to a pattern established and priced by the Board. If a maker could not or would not work to this pattern at the price, he was left out of the contract. It is likely that a certain amount of negotiation regarding both pattern and price still occurred during the early years of the new arrangement, but the balance had now been shifted strongly away from the Gunmakers' Company and towards the Board, who built up a cadre of well-informed people within their own organization and moved away from dependence on the Company, working directly with the individual makers rather than through the Company hierarchy. It is significant in this regard that this was the last occasion on which the term 'contract' was applied to an arrangement between the makers and the Board; henceforth a Royal Warrant was issued by the King as a general permission to create given numbers of components or for other services resulting in either parts of arms or complete arms.

1688 – 1715
BARRELS

Aldridge, George.	1713 Irish.
Banister, Thomas.	1699.
Blanckley, John.	1699.
Bleamire, Winifred.	1699.
Boulton, Peter.	1699; 1714 Irish.
Bye, Sarah.	1713 Irish.
Cole, Elias ✶	1710; 1715.
Cookes, Edward. ✶	1711.
Crips, Henry.	1699.
Dyer, Richard.	1699.
Dymond, Charles.	1699; 1713 Irish.
Farmer, Joseph. ✶	1708 – 1709.
Fisher, Mary.	1699.
Fort, Thomas.	1711; 1713 Irish.
Godward, Edward.	1699.
Goodby, James.	1713 Irish.
Hawkins, John Sr.	1713 Irish.
Hawley, Thomas.	1699.
Heasler, William. ✶	1699; 1713 Irish.
Hunt, Thomas.	1712.
Kipling, Charles.	1699; 1714 Irish.
Lawrence, Richard.	1707.
Loader, Richard.	1699.
Moore, William.	1699.
Nelson, John.	1699.
Nicholson, Edward Sr.	1699.
Nutt, William.	1699; 1713 Irish.
Peddall, James.	1699.
Phillips, Francis. ✶	1699; 1713 Irish.
Pickfatt, Humphrey Jr. ✶	1699; 1713 Irish.
Pike, Elizabeth.	1699.
Predden, William. ✶	1699.
Pye, Thomas.	1699.
Saunders, Thomas.	1699.
Silke, John.	1699.
Silke, Robert (I)	1699.
Smart, Francis. ✶	1699.
Sowerby, William. ✶	1714 Irish.
Spencer, Elizabeth.	1699.
Stebben, Robert.	1699.
Stringer, Ralph.	1699.
Tittensor, John. ✶	1707 – 1708.
Tittensor, William.	1708 – 1709.
Tough, Mary. ✶	1699; 1713 Irish.
Towle, Elizabeth.	1713 Irish.
Turvey, Edward. ✶	1699.
Turvey, John.	1699.
Vaughton, Humphrey.	1710.
Vaughton, Ryland.	1709.
Vaughton, Samuel.	1709 – 1711.
Ward, Thomas.	1699.
Watkinson, Mary.	1699.
Weston, Edmond. ✶	1699.
Weston, Richard.	1700.
White, Thomas.	1713 Irish.
Winesop, John.	1699.

LOCKS

Austin, Jacob.	1707 – 1709.
Bannister, Thomas.	1699; 1703.
Blanckley, John.	1699.
Bleamire, Winifred.	1699.
Boulton, Peter.	1699; 1714 Irish.
Brush, John.	1708 – 1709.
Bunday, John.	1713 Irish.
Bye, Richard.	1703.
Castle, Edmond.	1703.
Champante, Henry.	1702 – 1703.
Clifford, Peter.	1708 – 1709.
Cole, Elias. ✶	1710 – 1715.
Cookes, Edward. ✶	1707 – 1715.
Crips, Henry.	1699.

Dyer, Richard.	1699.	Smart, Francis.*	1699; 1708 – 1709.
Dymond, Charles.	1699; 1703; 1713 Irish.	Sowerby, William.*	1714 Irish.
		Spencer, Elizabeth.	1713 Irish.
Farmer, Joseph.*	1708 – 1715.	Stebben, Robert.	1699; 1703.
Fisher, Mary.	1699.	Stevens, Andrew.	1708 – 1709, 1711.
Fort, Mary.	1709; 1715.	Stringer, Ralph.	1699.
Fort, Thomas.	1691 – 1711.		
		Taylor, Godfrey	1699.
Godward, Edward.	1699.	Tough, Mary.*	1699.
Goodby, James.	1703 Sea Service.	Towle, Elizabeth.	1713 Irish.
Green, John.	1710 – 1711.	Turvey, Edward.*	1699; 1708.
Green, Thomas.*	1713 Irish.	Turvey, John.	1699.
Hawkins, John.*	1713 Irish.	Vaughan, John.*	1704; 1709; 1713; 1715 Irish;.
Hawkins, Richard.	1706.		
Hawley, Thomas.	1699; 1706.		
Heasler, William.*	1699.	Ward, Thomas.	1699; 1704.
Huggins, William.*	1704 – 1709.	Watkinson, Mary.	1699.
Hunt, Thomas.	1709.	Weston, Edmund.*	1699.
		Weston, Richard.	1700.
Kipling, Charles.	1699. 1714 Irish.	White, Thomas.*	1713 Irish.
Kipling, Richard.	1707.	Williams, John.*	1713 Irish; 1715.
		Winesop, John.*	1699; 1706.
Loader, Richard.	1699.		

BAYONETS

Austin, Thomas.	1703.
Mayo, John.*	1708.
Moore, Thomas.	1707; 1709.
Moore, William.	1699.
Bigglestone, Thomas.	1702 – 1711.
Nelson, John.	1699.
Chase, Samuel.	1703 – 1711.
Nicholson, Edward Sr.	1699.
Cole, Elias.*	1710 – 1715.
Nutt, William.	1699; 1713 Irish.
Crips, Henry.	1703 – 1704.
Peddell, James.	1699.
Hawgood, John.	1703 – 1708.
Phillips, Francis.*	1699.
Hawgood, Thomas.	1703 – 1711.
Pickfatt, Humphrey Jr.*	1699; 1703.
Pike, Elizabeth.	1699.
Mist, Phillip.	1703 – 1711.
Pollard, James.	1703.
Powell, Stephen.	1703.
Peddell, James.	1703.
Predden, William.*	1699.
Pye, Thomas.	1699.
Stone, Guy.	1708 – 1711.
Rose, William.*	1715.
Woodcraft, John.	1703 – 1711.

GUN FURNITURE

Saunders, Thomas.	1699; 1713 Irish.
Sibthorpe, Robert.	1713 Irish.
Bagley, Mathew. (B)	1715.
Silke, Robert (I).	1699.
Burgin, William. (B)	1715.

COMPLETE ARMS

(A terminal date of 1713 or 1714 indicates delivery of muskets for Irish service).

Aldridge, George.	1705 – 1712.
Arkill, Peter.	1696 – 1706.
Austin, Jacob.	1699 – 1709.
Austin, Mary.	1710 – 1711.
Austin, Thomas.	1688 – 1709.
Banks, Henry.	1688 – 1700.
Banister, Robert.	1702 – 1708.
Banister, Thomas.	1688 – 1705.
Batty, Henry.	1709 – 1710.
Bennett, John.	1691 – 1711.
Blanckley, John.	1688 – 1699.
Blanckley, Samuel.	1703 – 1706.
Bleamire, Winifred.	1688 – 1702.
Boulton, Peter.	1688 – 1714.
Bourne, William.	1700.
Bowers, George.	1704 – 1705.
Bridger, George.	1703 – 1707.
Brooke, John Sr.	1701 – 1710.
Brooke, Mary.	1688 – 1692.
Brush, John. *	1702 – 1713.
Bunday, John. *	1688 – 1711.
Bye, Richard.	1703 – 1710.
Bye, Sarah.	1710 – 1713.
Caddy, Edward.	1688 – 1707.
Carlyon, Nicholas.	1703 – 1709.
Castle, Edmund.	1701 – 1707.
Champante, Henry.	1702 – 1703.
Cole, Thomas.	1691 – 1694.
Crips, Henry.	1688 – 1706.
Crooke, Robert.	1691 – 1693.
Darke, Elizabeth.	1702 – 1709.
Darke, Harward.	1688 – 1696.
Darke, Thomas.	1709 – 1712.
Dolep, Andrew.	1703 – 1711.
Doughty, Thomas.	1696.
Dyer, Richard. *	1688 – 1712.
Dymond, Charles.	1692 – 1713.
Ebbutt, Lancelot. *	1704 – 1711.
Farmer, Joseph. *	1708 – 1710.
Finch, John.	1688 – 1690.
Fisher, George Jr.	1688 – 1693.
Fisher, Mary.	1693 – 1705.
Gale, Charles.	1688 – 1701.
Gale, Mary.	1702 – 1706.
Gardiner, Thomas.	1688 – 1693.
Gilbert, Ann.	1691 – 1696.
Godward, Edward.	1691 – 1709.
Goff, Benjamin.	1692 – 1711.
Goodby, James. *	1701 – 1712.
Green, Thomas. *	1697 – 1711.
Gregory, Richard.	1704 – 1707.
Groome, Collins.	1688 – 1690.
Groome, Margaret.	1690 – 1695.
Groome, Richard.	1688 – 1690.
Hall, John. *	1706 – 1712.
Harrison, Timothy.	1701 – 1711.
Hartwell, John.	1688 – 1691.
Hartwell, Susannah.	1689 – 1690.
Harvey, Robert.	1703 – 1705.
Hawden, Flower.	1706 – 1711.
Hawkins, John. *	1688 – 1711.
Hawkins, Richard.	1701 – 1708.
Hawley, Joseph.	1703 – 1705.
Hawley, Thomas.	1688 – 1706.
Haynes, James.	1706 – 1711.
Heasler, William. *	1692 – 1713.
Hodgson, Elizabeth.	1689 – 1691.
Hodgson, Paul.	1690 – 1693.
Hodgson, Thomas.	1688 – 1693.
Huggins, William. *	1706 – 1711.
Hunt, Thomas.	1708 – 1712.
Jackson, Christopher. *	1709 – 1711.
Johnson, Christopher. *	1708 – 1713.
Johnson, John. *	1696 – 1713.
Kipling, Charles. *	1690 – 1712.
Kipling, Richard. *	1701 –1711.
Kirby, Henry.	1689 – 1706.
Lawrence, Richard.	1689 1711.
Loader, Edward. *	1707 – 1712.
Loader, Richard.	1689 – 1711.

COMPONENTS & PRODUCTION CONTRACTORS

Looker, Robert (I)	1688 – 1689.	Smart, Francis. ✷	1695 – 1712.
Luther, Stephen.	1709.	Smith, William.	1708.
		Smithett, George Sr. ✷	1703 - 1711.
Mayo, John. ✷	1701 – 1710.	Sowerby, William. ✷	1708 – 1712.
Moore, Thomas.	1700 – 1711.	Spencer, James.	1689 – 1696.
Moore, William. ✷	1689 – 1711.	Stace, Joseph.	1688 – 1689.
Morris, Anthony.	1692 – 1696.	Stebben, Robert.	1695 – 1710.
Morris, Isaac. ✷	1701 – 1712.	Stevens, Andrew.	1708 – 1711.
		Stringer, Ralph.	1695 – 1711.
Neave, Richard.	1703.		
Nelson, John.	1690 – 1702.	Taylor, George.	1701 – 1711.
Nicholson, Edward, Sr. ✷	1688 – 1711.	Taylor, Godfrey.	1688 – 1699.
Nicholson, Edward, Jr. ✷	1706 – 1710.	Taylor, Thomas.	1710.
Nutt, William. ✷	1689 – 1712.	Thomas, Ralph.	1696 – 1702.
		Tittensor, John. ✷	1707 – 1711.
Oldfield, Henry.	1696.	Tittensor, Joseph.	1710.
		Tittensor, William.	1710 – 1711.
Peddell, James. ✷	1688 – 1715.	Tough, Mary. ✷`	1694 – 1713.
Pettyman, Joseph.	1706.	Tough, Robert. ✷	1705 – 1711.
Phillips, Francis. ✷	1689 – 1713.	Towle, Elizabeth. ✷	1693 – 1713.
Phillips, Thomas. ✷	1708 – 1711.	Towle, Thomas (I).	1688 – 1691.
Pickfatt, Humphrey, Sr.	1688 – 1696.	Towle, Thomas (II). ✷	1703 – 1711.
Pickfatt, Humphrey, Jr. ✷	1689 – 1713.	Trulock, George.	1695 – 1704.
Pike, Christopher.	1688 – 1695.	Turner, John.	1695.
Pitts, Mordecai.	1704 – 1709.	Turvey, Edward. ✷	1690 – 1713.
Pollard, James.	1702 – 1705.	Turvey, John.	1690 – 1701.
Powell, Widow Martha. ✷	1714 Irish muskets.	Tylor, Thomas. ✷	1688 – 1711.
Powell, Stephen. ✷	1696 – 1711.	Vaughan, John. ✷	1706 – 1712.
Powell, Thomas.	1701 – 1711.	Vaughton, Samuel.	1709 – 1711.
Predden, William. ✷	1691 – 1714.		
Press, Edward.	1703 – 1705.	Wagg, Francis (I).	1688 – 1690.
Probin, Thomas. ✷	1708 – 1711.	Wagg, Francis (II).	1693 – 1696.
Pye, Humphrey.	1688 – 1694.	Ward, Thomas.	1695 – 1706.
Pye, Thomas.	1693 – 1706.	Warren, Charles.	1696.
		Watkinson, John.	1688 – 1693.
Radbird, James.	1703.	Watkinson, Mary.	1695 – 1705.
Rose, William. ✷	1706 – 1711.	Weston, Edmond.	1690 – 1711.
		Weston, Richard.	1700 – 1710.
Saunders, Thomas. ✷	1688 – 1711.	White, Thomas. ✷	1702 – 1711.
Savage, Richard.	1688 – 1692.	Williams, John. ✷	1701 – 1713.
Shorey, Joseph.	1702 – 1711.	Willowes, John. ✷	1705.
Sibley, John.	1688 – 1711.	Winesop, John. ✷	1696 – 1706.
Sibthorpe. Robert. ✷	1706 – 1713.		
Silke, John.	1689 – 1705.		
Silke, Robert (I).	1688 – 1700.		
Silke, Robert (II).	1694 – 1695.		

1714 – 1717

ROUGH STOCKERS & SETTERS UP, beginning with the contract of 15 Sept. 1714 with bill dates from Dec. 1714, but not including the commencement of the Pattern 1718 arms. This interim period represents the functional beginning of the 'Ordnance System' of production.

Name	Dates
Aldridge, George.✶	1714 – 1716.
Brazier, William.	1714 – 1716.
Brooke, John, Sr.✶	1715.
Brooke, John, Jr.	1715 – 1718.
Drew, John.✶	1715 – 1716.
Dyer, Richard.✶	1715 – 1715.
Ebbutt, Lancelot.✶	1715 – 1716.
Fort, Mary.✶	1715.
Goodby, James.✶	1715 – 1716.
Green, Thomas.✶	1715 – 1716.
Hall, John.✶	1715 – 1716.
Harrison, Timothy.✶	1715 – 1716.
Hawkins, John Sr.✶	1714 – 1716.
Hawkins, John, Jr.	1715.
Haynes, James.✶	1715 – 1716.
Heasler, William.✶	1715 – 1717.
Jackson, Christopher.✶	1715.
Johnson, Christopher.✶	1715 – 1716.
Johnson, John.✶	1715 – 1717.
Kipling, Charles.✶	1715.
Kipling, Richard.✶	1715.
Kirkham, Henry.✶	1716.
Lawrence, Margaret.	1715.
Lloyd, Evan.✶	1716 – 1717.
Loader, Edward.✶	1715 – 1717.
Loader, Richard.✶	1714 – 1715.
Looker, Robert (II).	1715 – 1717.
Moore, William.✶	1715 – 1716.
Morris, Isaac.✶	1715 – 1717.
Nutt, William.✶	1715 – 1716.
Peddell, James.✶	1715 – 1716.
Phillips, Francis.✶	1715.
Phillips, Thomas.✶	1715.
Pickfatt, Humphrey, Jr.	1715 – 1717.
Powell, Hugh.✶	1716 – 1717, Rough Stocker only.
Powell, Widow Martha.✶	1715.
Powell, Stephen.✶	1715.
Predden, William.✶	1715 – 1716.
Pringle, John.✶	1715 – 1717.
Probin, Thomas.✶	1717.
Rose, William.✶	1715.
Sale, Edward.	1715.
Saunders, Thomas.✶	1714 – 1716.
Sibthorpe, Robert.✶	1715 – 1717.
Sinckler, Richard.✶	1715 – 1716.
Smart, Francis.✶	1715 – 1716.
Smithett, George, Sr.✶	1715 – 1716.
Smithett, George, Jr.	1716.
Sowerby, Joseph.✶	1716.
Sowerby, William.✶	1714 – 1716.
Stringer, Elizabeth.	1715.
Tittensor, John.✶	1715.
Tough, Mary.✶	1714 – 1716.
Tough, Robert.✶	1715 – 1716.
Towle, Elizabeth.✶	1714 – 1716.
Towle, Thomas (II).✶	1714 – 1716.
Turvey, Edward.✶	1715 – 1716.
Tylor, Thomas.✶	1715.
Vaughan, John.✶	1714 – 1716.
Welford, Richard.✶	1715.
White, John.	1716 – 1717.
White, Thomas.✶	1715 – 1716.
Williams, John.✶	1715 – 1716.
Winesop, John.✶	1715 – 1717.

1717 – 1725

This period is characterised by the production of the 'Pattern of the 10,000' or Pattern 1718 series with flat plain (i.e. with neither tumbler- nor pan-bridle) locks and iron-furniture, through 1722. This is the last pattern to be produced by a large number of contractors each supplying small amounts of a certain pattern. In 1722, Lewis Barber delivered several pattern arms (none of which were subsequently produced) and several gunmakers delivered complete groups of regimental arms. Iron ramrods were tentatively introduced in 1724 on a set of arms for regiments serving in Ireland.

BARRELS
Cole, Elias. ✶	1718 – 1723.

LOCKS
Cole, Elias. ✶	1718 – 1720.
Cookes, Edward. ✶	1718 – 1719.
Farlow, John. ✶	1719 – 1720.
Farmer, Joseph. ✶	1718.
Mayo, John. ✶	1719 – 1720.
Rose, William. ✶	1720.
Vaughan, John. ✶	1718 – 1720.
Williams, John. ✶	1719 – 1720.

ENGRAVERS
Banbury, William.	1719.
Caslon, William.	1717 – 1719.

FURNITURE, Iron, Brass
Burgin, Mary. ✶ (B)	1718 – 1719.
Hollier, Thomas. ✶ (I,B)	1716 – 1718.
Probin, Thomas. ✶ (I)	1717 – 1720.
Weston, Edward. ✶ (I)	1717.

BAYONETS
Cole, Elias. ✶	1719 – 1722.
Hollier, Thomas. ✶	1718 – 1726.

RAMMERS, Wooden
Thompson, John.	1715 – 1723.

ROUGH STOCKERS
Davis, Robert.	1719 – 1720.
Everard, William.	1718 – 1719.
Frankland, Richard.	1720.
Johnson, Margaret. ✶	1719 – 1720.
Loyd, Evan.	1718 – 1721.
Powell, Hugh.	1718 – 1719.
Powell, Richard.	1718 – 1719.
Sinckler, Ralph.	1718 – 1719.
Vaughan, John. ✶	1719 – 1720.
Waller, Richard. ✶	1718 – 1720.
Ward, Richard.	1718 – 1719.
White, John.	1719.
Woodell, George.	1719.

STOCKERS & SETTERS UP
Barnes, John.	1719 – 1720.
Barras, Ralph.	1718 – 1720.
Buckmaster, Robert.	1719 – 1720.
Buckmaster, William.	1720.
Brooke, John, Jr.	1718.
Bunday, John.	1718.
Carter, Benjamin.	1718 – 1721.
Collet, Joseph.	1718 – 1721.
Davis, William. ✶	1718 – 1719.
Dennison, John.	1718 – 1720.
Drew, John. ✶	1719 – 1720.
Fitchett, William.	1719.
Halfhide, George.	1719 – 1720.
Hall, John. ✶	1718 – 1720.
Johnson, John. ✶	1719 – 1720.
Johnson, Margaret. ✶	1718 – 1720.
Kirkham, Henry.	1718 – 1721.
Loyd, Evan. ✶	1718 – 1721.
Loader, Mary.	1719 – 1720.
Looker, Robert (II).	1718.

Moore, William. ✱	1719 – 1720.	Tough, Mary. ✱	1718.
Morris, Isaac. ✱	1719 – 1721.	Tough Robert. ✱	1718 – 1721.
Phillips, Francis. ✱	1719 – 1720	Towle, Thomas (II). ✱	1718 – 1720.
Phillips, Thomas. ✱	1720 – 1721.	Turvey, William.	1721.
Pickfatt, Humphrey. ✱	1720.	Vaughan, John. ✱	1719 – 1720.
Powell, Hugh. ✱	1719 – 1720.	Welford, Richard. ✱	1719 – 1720.
Predden, William. ✱	1719 – 1720.	White, John.	1718 – 1722.
Press, Joan.	1719 – 1720.	Williams, John.	1719 – 1720.
Pringle, John. ✱	1718 – 1721.	Willowes, John.	1719 – 1720.
Rose, William. ✱	1720.		
Simpson, William.	1718 – 1720.		

COMPLETE ARMS

Barber, Lewis. ✱	1722, 1725.
Clarkson, Joseph. ✱	1723.
Farmer, Joseph. ✱	1723.
Freeman, James. ✱	1723.
Phillips, Thomas. ✱	1724.
Sinckler, Richard. ✱	1724.

Sinckler, Ralph. ✱	1718 – 1720.
Smart, Francis. ✱	1719 – 1721.
Smart, John.	1720.
Smithett, Dorothy.	1719 – 1720.
Smithett, George Sr. & Jr.	1718 – 1720.
Sowerby, Joseph. ✱	1718 – 1720.
Sowerby, William. ✱	1719 – 1720.
Tittensor, John. ✱	1718.

1726 – 1738

At the beginning of this period there was a stockpiling of components for a new series of arms for the first time designated as the King's Pattern, which, in 1730, commenced production with the Pattern 1730 Long Land Musket, Carbine and Pistol.

BARRELS

Cole, Elias. ✱	1726 – 1734.
Cookes, Edward. ✱	1727 – 1738.
Farmer, Joseph. ✱	1727 – 1732.
Jordan, Edward. ✱	1733 – 1738.
Wooldridge, Richard. ✱	1727

LOCKS

Cole, Elias. ✱	1728 – 1730.
Cookes, Edward. ✱	1729.
Farlow, John. ✱	1728 – 1730; 1737.
Farmer, Joseph. ✱	1728 – 1731.
Jordan, Edward. ✱	1734 – 1738.
Smith, John. ✱	1736 – 1738.
Vaughan, John. ✱	1728 – 1738.

BAYONETS

Hollier, Thomas. ✱	1726 – 1732; 1738.
Huggins, William. ✱	1727 – 1730.

FURNITURE, Brass

Burgin, Mary. ✱	1727.
Hollier, Thomas. ✱	1727 – 1738, also iron to 1730.
Huggins, William. ✱	1728 – 1730, also iron.

RAMMERS, Wood.

Green, Mary.	1728.
Green, Thomas. ✱	1727 – 1728.
Haynes, Jeremiah.	1733.

RAMMERS, Steel.

Hollier, Thomas. ✱	1728, 1735.

ROUGH STOCKERS

Waller, Richard. ✱	1730 – 1738.

COMPONENTS & PRODUCTION CONTRACTORS

SETTERS UP		COMPLETE ARMS	
Barber, Lewis. *	1730 – 1738.	Barber, Lewis. *	1727.
Green, Thomas. *	1728.	Clarkson, Joseph. *	1728.
Johnson, Margaret. *	1729.	Farmer, Joseph. *	1728.
Kirkham, Henry. *	1728.	Freeman, James. *	1727.
Lampheere, Richard.	1728.	Probin, Thomas. *	1727.
Lloyd, Evan. *	1728.	Sinckler, Richard. *	1727 – 1728.
Pickfatt, Charles. *	1727 – 1738.	Sowerby, William. *	1728.
Wooldridge, Richard. *	1735.	Wooldridge, Richard. *	1727.

1739 – 1750

Production during this period was governed by war requirements. All major components were simplified in design at the beginning of the period. Eight new setters up were introduced into the system, beginning in 1742. With the exception of Jordan & Farmer, all complete arms delivered during this period were small quantity and/or special order weapons.

BARRELS
Clarke, William.	1740 – 1742
(repairing barrels to 1749; bullet moulds).	
Cookes, Edward. *	1738 – 1741.
Farmer, James. *	1741 – 1751.
Jordan, Edward. *	1738 – 1748.
Jordan, Thomas. *	1745.
Jordan & Farmer.	1741 – 1745.
(Chiefly COMPLETE arms).	
Tippin & Edge.	1748.
Willet, John. *	1743 – 1748.
Wooldridge, Richard. *	1742

LOCKS
Clarke, William.	1741 – 1742.
Farmer, James. *	1741 – 1751.
Jordan, Edward. *	1738 – 1750.
Jordan & Farmer.	1742 – 1744.
Smith, John. *	1739 – 1748.
Tippin & Edge.	1748.
Vaughan, John. *	1739 – 1747.
Willet, John. *	1743 – 1748.
Wood, John. *	1748.

FURNITURE, Brass
Hollier, Thomas. *	1739 – 1752.

BAYONETS
Hollier, Thomas. *	1739 – 1753.

ENGRAVERS
Sharp, Richard.	1743 – 1755.

RAMMERS, Steel
Hollier, Thomas. *	1748 – 1749.

RAMMERS, Wooden.
Meakin, Nathaniel. *	1745 – 1752.

ROUGH STOCKERS
Waller, Richard. *	1739 – 1750.

SETTERS UP
Barber, James. *	1741 – 1749.
Barber, Lewis. *	1739 – 1741.
Birkell, William.	1742 – 1749.
Gandon, Peter. *	1743 – 1749.
Gluvias, Hewit.	1743 – 1744.
Hirst, John. *	1746 – 1749.
Pickfatt, Charles. *	1739 – 1749.
Sale, Edward.	1743 – 1749.
Taylor, John.	1747 – 1749.
Trevey, Nathaniel.	1743 – 1749.
Wilson, Richard. *	1746 – 1749.

COMPLETE ARMS

Barber, James. *	1741 – 1749.	Hatcher, Thomas.	1745, 1750, 1752.
Barber, Lewis. *	1738 – 1741.	Jordan & Farmer.	1742 – 1746.
Farmer, James. *	1741 – 1748.	Tippin & Edge.	1747.
Farmer, Joseph. *	1745 – 1746.	Wooldridge, Richard. *	1742 – 1743.
Grice, John & William.	1742.		

1754 – 1774

During this period, setting up came under the almost complete control of John Hirst. The Ordnance System achieved peak results during the Seven Years' (French & Indian) War 1755-1763.

BARRELS

Edge & Son, Richard.	1757 – 1774.	Jordan, Thomas. *	1757 –1762.
Farmer, James. *	1757 – 1759.	Molineux, Richard.	1759.
Farmer & Galton.	1758 – 1774.	Newton, Edward.	1759 – 1761.
Galton, Samuel. *	1756 – 1759.	Perry, William.	1756 – 1757.
Grice, Joseph (I)	1762 – 1771.	Stamps, Thomas.	1767.
Grice Joseph (II). *	1773 – 1774.	Tomkys & Short.	1757 – 1768.
Grice & Edge.	1757.	Vernon, George.	1757 – 1758.
Grice & Son, William. *	1760 – 1771.	Whately, John. *	1762 – 1764.
Harris, Joseph.	1757 – 1762.	Willet, John. *	1757 – 1758.
Harris & Barker. *	1762 – 1774.	Willetts, Benjamin. *	1769 – 1771.
Jordan, Edward. *	1755 – 1758.	Wilson, Richard. *	1756 – 1761.
Jordan, Thomas. *	1757 – 1762.	Wood, John. *	1756 – 1757.
Molineux, Richard.	1759.		
Newton, Edward.	1759 – 1761.	### FURNITURE, Brass	
Oughton, Joseph (I)	1755 – 1759.	Hartwell & Mayor.	1755 – 1767.
Oughton & Son. Joseph. *	1759 – 1774.	Jordan, Thomas.	1757 – 1762.
Tomkys & Short.	1758 – 1768.	Mayor, Jane. *	1768 – 1775.
Whately, John. *	1757 – 1774.	Mayor, Thomas.	1761 – 1767.
Willet, John, *	1757 – 1758.		
Willetts, Benjamin. *	1769 – 1774.	### BAYONETS	
Wilson, Richard. *	1756 – 1761.	Loxham, William & Edward. *	1754 – 1774.

LOCKS

Edge & Son, Richard.	1760 – 1775.	### RAMMERS, Steel	
Farmer, James. *	1757 – 1759.	Edge & Son, Richard.	1762 – 1774.
Farmer & Galton.	1757 – 1774.	Grice & Son, William. *	1761 – 1775.
Galton, Samuel. *	1757.	Sharp, William. *	1754 – 1775.
Grice, Joseph. *	1764.		
Grice & Edge.	1756 – 1759.	### RAMMERS, Wooden	
Grice & Son, William. *	1760 – 1770.	Meakin, Nathaniel. *	1752 – 1755.
Harris, Joseph.	1757 – 1758.	Freeman, Samuel. *	1756 – 1775.
Haskins, George.	1757 - 1759; 1767 - 1769.		
Haskins & Vernon.	1758 – 1766.	### ENGRAVERS	
Jordan, Edward. *	1755 – 1758.	Sharp, Richard. *	1754 – 1755.
		Sharp, William. *	1755 – 1774.

ROUGH STOCKERS

Kelso, Isaac.	1756 – 1757.
Loder, Joseph.✶	1756; 1769 – 1774.
Waller, James.✶	1769 – 1782.
Waller, Richard & James.	1755 – 1769.

SETTERS UP

Brazier, John.	1756 – 1757.
Buckmaster, John.	1756 – 1757.
Bumford, John.	1756 – 1757.
Gandon, Peter.✶	1756.
Hadley, Henry.	1757.
Hall, Joseph.	1756 – 1757.
Henshaw, Thomas.	1757.
Hirst, John.✶	1755 – 1774.
How, Richard.	1756 – 1758.
Memory, Michael.✶	1756 – 1758.
Newton, Edward.	1759 – 1761.
Pickfatt, Charles.✶	1756 – 1757.
Wilson, Richard.✶	1756 – 1761.

COMPLETE ARMS

Bumford, John.	1771.
Farmer, James.✶	1756 – 1759.
Jordan, Edward.✶	1756 – 1758.
Jordan, Thomas.✶	1757 – 1762.
Newton, Edward.	1759 – 1761.

1775 – 1783

During this period production is centred on the demands of the American War, and the Ordnance System begins to weaken from 1778. Three virtual monopolies are broken as additional firms are taken on to expand the production: Waller and Loder for rough stocking, Hirst for setting up, and Loxhams for bayonets. A large proportion of Sea Service muskets is allowed to be supplied complete, and several thousand Short Land muskets are also delivered complete by Pratt and Hirst. Almost 100,000 muskets are produced to Ordnance pattern in Liège and imported.

BARRELS

Barker, Matthias.	1775 – 1782.
Bissell, Isaac.	1779 – 1783.
Falkner & Co., Edward.	1778 – 1779.
Galton & Son, Samuel.✶	1775 – 1782.
Green, Thomas & Hezekiah.✶	1780 – 1782.
Grice, Joseph (II)✶	1773 – 1783.
Hadley, Thomas.✶	1776 – 1781.
Harris & Barker.✶	1774 – 1775.
Holden, William.✶	1776 – 1783.
Oughton, Joseph (II).✶	1776 – 1783.
Oughton & Son, Joseph.✶	1775 – 1776.
Pratt, John.	1779 – 1780.
Whately, John (II)✶	1776 – 1783.
Willetts, Benjamin.✶	1775 – 1782.

LOCKS

Blakemore, Thomas.✶	1775 – 1781.
Falkner & Co., Edward.	1778 – 1779.
Galton & Son, Samuel.✶	1775 – 1782.
Green, Thomas & Hezekiah.✶	1780.
Grice & Son, William.✶	1775 – 1780.
Hadley, Thomas.✶	1776 – 1781.
Hartwell & Haskins.	1775.
Holden, William.	1776 – 1782.
Leonard, John.	1780; 1782.
Nock, Henry.✶	1776 – 1781.
Oughton, Joseph.✶	1776 – 1782.
Whately, John.	1776, 1778 – 1779.
Whately & Son, John.✶	1775.
Willetts, Benjamin.✶	1775 – 1780.

BAYONETS

Dawes, Samuel.✶	1779 – 1782.
Falkner & Co., Edward.	1778 – 1779.
Galton & Son, Samuel.✶	1778 – 1782.
Gill, Thomas.✶	1778 – 1781.
Hadley, Thomas.	1780 – 1781.
Harvey, George & Samuel.	1779 – 1781.
Hornbuckle, Richard.	1778 – 1780.
Horsley, John.	1778 – 1780.
Loxham, William & Edward.✶	1775 – 1780.
Pardoe, Ambrose (Liège).	1780.
Pratt, John.	1779 – 1780.

ENGRAVERS

Sharp, William.✶	1775 – 1781.
Sharp & Son, William.✶	1781 – 1782.

FURNITURE

Mayor, Jane. ✱	1775 – 1782.

RAMMERS, Steel

Falkner & Co., Edward.	1778 – 1779.
Galton & Son, Samuel. ✱	1775 – 1782.
Grice & Son, William. ✱	1775 – 1782.
Hadley, Thomas. ✱	1776 – 1781.
Sharp, William.	1774 – 1782.

RAMMERS, Wooden

Freeman, Samuel. ✱	1775 – 1781.
Sharp, William. ✱	1782.

ROUGH STOCKERS

Hunt, Joseph.	1778 – 1781.
Loder, Joseph. ✱	1769 – 1782.
Memory, Michael. ✱	1778 – 1782.
Nicholson, William. ✱	1782.
Ross, Robert.	1778 – 1779.
Trested, Richard. ✱	1777 – 1783.
Tucker, Thomas. ✱	1777 – 1783.
Waller, James. ✱	1769 – 1781.

SETTERS UP

Bate, Edward.	1779 – 1782.
Davies, Thompson.	1777 – 1780.
Davidson, Alexander.	1781 – 1783.
Goff, Daniel. ✱	1782 – 1783.
Harrison, John.	1778 – 1779.
Harrison & Thompson. ✱	1779 – 1783.
Hirst, John & James.	1774 – 1777.
Hirst, James. ✱	1777 – 1783.
Humphreys, Veritas.	1778 – 1781.
Hunt, Joseph.	1778 – 1781.
Nicholson, William. ✱	1782.
Pratt, John.	1777 – 1781.
Ross, Robert.	1778 – 1779.
Wilson, William.	1777 – 1782.

COMPLETE ARMS

Barker, Matthias. (Rifles)	1776
Bate, Edward.	1779 – 1782.
Galton & Son, Samuel.(Rifles)	1776
Grice & Son, William.(Rifles)	1776
Harrison, John.	1778 – 1779.
Harrison & Thompson. ✱	1778 – 1783.
Hirst, James.	1778 – 1782.
Memory, Michael. ✱	1778 – 1782.
Moore, Daniel.	1778 – 1780.
Nock, Henry. ✱ (7–bbl guns)	1779 – 1781.
Pratt, John.	1778 – 1780.
Willetts, Benjamin. (Rifles)	1776

1784 – 1792

With the return of peace tens of thousands of muskets and other arms are sold off, largely emptying the stores to a minimum level. Experimentation with breech-loading and rifled arms, and the re-design of the infantry musket are the chief concerns of the Master General, and there is only a small production of regulation carbines during the period.

BARRELS

Galton & Son, Samuel. ✱	1787.
Goff, Daniel. ✱	1785.
Green, Thomas & Hezekiah. ✱	1786; 1788.
Hennem, Jonathan. ✱	1790 – 1792.
Holden, William. ✱	1786; 1788.
Nock, Henry. ✱	1790 – 1792.
Oughton, Joseph (II). ✱	1787 – 1788.
Whately, John (II). ✱	1785 – 1788.

LOCKS

Grice & Son, William. ✱	1789 – 1790.
Hennem, Jonathan. ✱	1784 – 1792.
Nock, Henry. ✱	1783 – 1792.

BAYONETS

Gill, Thomas. ✱	1789 – 1792.

COMPLETE ARMS
(except certain components)

Egg, Durs.★	1783; 1786.
Hennem, Jonathan.★	1785 – 1792.
Nock, Henry.★	1784 – 1792.

ENGRAVERS

Gyde, John.★	1786 – 1792.
Sharp & Son. William.★	1784 – 1786.
Witton, Joseph.★	1786 – 1792.

BRASS FURNITURE

Mayor, Jane.★	1785 – 1792.

ROUGH STOCKERS

Loder, Joseph.★	1785.
Trested, Richard.★	1784 – 1786.
Tucker, Thomas.★	1785.

SETTERS UP

Goff, Daniel.★	1785.
Hennem, Jonathan.★	1785 – 1792.
Hirst, James.	1784.
Nock, Henry.★	1784 – 1792.

1793 – 1803

The outbreak of war against France in 1793 found the Ordnance unprepared and its stores wholly unable to meet the demands of regular, militia, volunteer and hired foreign troops. Orders were placed in Liège, but the principality was occupied by the French in 1795; purchases of arms from the East India Company, beginning in 1794, helped to fill the gap, and led to the adoption of the cheaper and simpler 'India Pattern' as the principal infantry musket design for the period from 1795 to 1815. London, Birmingham and Bristol merchants and contractors supplied quantities of non-regulation pattern military arms during the period to 1798, with Nock, Egg and Barnett largely supplying the Ordnance with regulation arms during these years. By 1797 a system had evolved with both London and Birmingham contractors supplying a variety of arms, with the India Pattern musket as the principal product.

BARRELS

Archer, Thomas.	1794 – 1809.	Holden, William.★	1794 – 1799.
Baker, Ezekiel.	1794 – 1804.	Limbrey, Matthew.	1797.
Bayliss, Joseph.★	1798 – 1803.	Marwood, William.★	1798 – 1802.
Bird, William.	1790 – 1795.	Nock, Henry.★	1793 – 1802.
Blair, David.	1793 – 1802.	Oughton, John.	1796 – 1802.
Bonus & Holbrook.	1796.	Oughton, Joseph (II).★	1794.
Dawes, William & Samuel.★	1798 – 1803.	Oughton & Parr.	1794 – 1796.
Edge, Thomas.	1799 – 1801.	Parsons, Gad.★	1798 – 1802.
Galton & Son, Samuel.★	1793 – 1796.	Portlock, Thomas.★	1798 – 1802.
Galton Jr., Samuel.★	1796 – 1803.	Reynolds, Thomas.★	1796 – 1800.
Gill, John.	1801 – 1802.	Rock, William.★	1799 – 1802.
Gill, Thomas.★	1793 – 1801.	Russell, John.★	1794 – 1802.
Gill, Thomas & James.	1801.	Russell, Thomas.★	1793 – 1802.
Green, Thomas & Hezekiah.★	1794.	Simes, John.★	1800 – 1802.
Green & Price.	1794 – 1804.	Stokes, Thomas.★	1798 – 1802.
Grice, Joseph (II).★	1794 – 1802.	Taylor & Davies.	1799 – 1802.
Harrison & Thompson.★	1793 – 1802.	Truman, Elizabeth.★	1801 – 1802.
Heely & Co., John & Samuel.★	1794 – 1802.	Truman, John.	1798 – 1801.

Whately, John. ✱	1794 – 1802.
Wheeler, Robert Jr. ✱	1800 – 1802.
Willetts, Benjamin. ✱	1794 – 1795.
Willetts, Mary.	1797 – 1802.
Willetts & Holden. ✱	1797.

LOCKS

Bird, William.	1790 – 1795.
Blair, David.	1793 – 1802.
Blakemore, Thomas. ✱	1793 – 1802.
Bonus & Holbrook.	1796.
Dick, Walter. ✱	1796 – 1803.
Gill, John.	1801 – 1802.
Gill, Thomas. ✱	1793 – 1801.
Gill, Thomas & James.	1801.
Grice, Joseph (II). ✱	1794 – 1802.
Grice & Morris.	1797 – 1800.
Harrison & Thompson. ✱	1793 – 1802.
Ketland & Walker. ✱	1796 – 1802.
Limbrey, Matthew.	1797.
Marwood, William. ✱	1798 – 1802.
Negus, Ann.	1802.
Negus, James.	1794 – 1802.
Nock, Henry. ✱	1793 – 1802.
Pritchett, Samuel. ✱	1797, 1802.
Reynolds, Thomas. ✱	1796.
Rock, William. ✱	1799 – 1802.
Round, Joseph. ✱	1801 – 1802.
Sherwood, Joseph Sr. ✱	1796 – 1800.
Smith, James. ✱	1793 – 1802.
Spittle, Peter. ✱	1801 – 1802.
Whately, John. ✱	1794 – 1802.
Wheeler, Robert Jr. ✱	1797 – 1800.
Whitehead, Francis. ✱	1801 – 1802.
Wilkes, James.	1799 – 1800.
Willetts, Benjamin. ✱	1794 – 1795.
Willetts, Mary.	1797 – 1802.
Willetts & Holden. ✱	1797.
Wood, Joseph.	1800 – 1801.

BAYONETS

Amphlett, Richard.	1795 – 1796.
Atherton, John.	1795.
Barnett, Thomas. ✱	1797.
Cam, James.	1795 – 1796.
Dawes, Samuel.	1794 – 1803.
Dawes, William & Samuel.	1798 – 1802.
Gill, Thomas. ✱	1793 – 1801.
Makin, James. ✱	1795 – 1802.
Osborne, Henry. ✱	1797 – 1802.
Oughton, John.	1796 – 1803.
Rock, William. ✱	1799 – 1802.
Wheeler, Robert Jr. ✱	1797 – 1802.
Woolley, James. ✱	1799 – 1802.

GUN FURNITURE

Mayor, Jane. ✱	1793 – 1796.
Mayor, Joseph. ✱	1797 – 1803.
Warner & Son.	1797 – 1801.

IMPLEMENTS (Rifle buttbox)

Baker, Ezekiel. ✱	1800.
Nock, Henry. ✱	1800.
Sherwood, Joseph Sr. ✱	1801.

ROUGH STOCKERS & SETTERS UP

Barnett, Thomas. ✱	1793 – 1803.
Knubley, John.	1793 – 1794.
Limbrey, Matthew.	1794 – 1797.
Marwood, William. ✱	1798 – 1802.
Memory & Wright.	1793 – 1802.
Nock, Henry. ✱	1793 – 1802.
Pritchett, Samuel. ✱	1793 – 1802.
Raby, Alexander. Merchant.	1800 – 1803.
Reynolds, Thomas. ✱	1796 – 1802.

SETTERS UP

Birkley, Thomas.	1794.
Memory, Michael.	1793.
Nicholson, Ann.	1794.
Nicholson, William. ✱	1795; 1796.

COMPLETE ARMS

Adam, Alexander. Merchant.	1798.
Barnett, Thomas.	1793 – 1803.
Blair & Sutherland. ✱	1804.–
Bonus & Holbrook.	1796.
Brander, Martin. ✱	1793 – 1803.
Egg, Durs. ✱	1793 – 1802.
Galton & Son, Samuel. ✱	1793 – 1796.
Galton Jr., Samuel.	1796 – 1802.
Gill, John.	1801 – 1802.
Gill, Thomas. ✱	1793 – 1801.
Gill, Thomas & James.	1801.
Gill, Thomas, James & John.	1801.

Grice, Joseph (II). ✱	1797 – 1802.	Rea, John. ✱	1793 – 1802.
Grice & Morris.	1797 – 1800.	Reynolds, Thomas. ✱	1795 – 1802.
Harrison & Thompson. ✱	1793 – 1802.	Sherwood, Joseph Sr. ✱	1797 – 1799.
Ketland & Walker. ✱	1796 – 1802.	Thompson, James. ✱	1802.
Knubley, John.	1793 – 1794.	Whately, John. ✱	1794 – 1802.
Limbrey, Matthew.	1794 – 1797.	Wheeler, Robert Jr. ✱	1797 – 1802.
Marwood, William. ✱	1798 – 1802.	Wilkes, James.	1799 – 1800.
Memory & Wright.	1793 – 1802.	Willetts, Mary.	1797 – 1802.
Nock, Henry. ✱	1793 – 1802.	Wright, Robert. ✱	1797 – 1801.
Pritchett, Samuel. ✱	1793 – 1802.	Wright & Co., Robert. ✱	1801 – 1802.
Raby, Alexander. Merchant.	1794 – 1803.		

1803 – 1819

During the 'pause' in the war during the Peace of Amiens (27 Mar. 1802 - 16 May 1803), the Ordnance attempted to return to production of a higher quality infantry musket, the New Land Pattern, and some components were delivered. The limitations of the contractor system were recognized and an additional fabrication centre was established with the skilled workmen of the Small Gun Office in the Tower of London as the core of an expanded establishment, the Royal Manufactory of Small Arms. But its personnel were still dependent upon the London and Birmingham trades for their components. An entirely new system based on open contracts was set up in March 1804, whereby the London and Birmingham contractors supplied both components for their own production of complete arms and components for the Royal Manufactory. The new system brought a change in book-keeping methods which eliminated the separate listing of components from early 1805. We have only monthly lists of (Birmingham) contractors' names and the amounts of money paid to them. This system continued in effect until mid-1814 when a false halt was made in India Pattern musket production after the occupation of Paris. A year later it resumed for several months, and finally ceased in the autumn of 1815. Steady, but limited, production of the New Land Pattern arms continued until mid-1817. Some heavy dragoon carbines and a few cadet carbines closed production early in 1819.

BARRELS

Archer & Son, Thomas.	1805 – 1808.	Deakin, William. ✱	1805 – 1810.
Arnold, Francis. Rifler.	1806 – 1811.	Deakin & Son, William.	1810 – 1818.
Ashton, Thomas.	1804 – 1818.	Galton Jr., Samuel. ✱	1803 – 1819.
Aspinall, James.	1812.	Gill, Elizabeth. ✱	1817 – 1819.
Bayliss, Joseph.	1803 – 1811.	Gill & Co., John. ✱	1805 – 1817.
Bayliss & Son, Joseph.	1811 – 1818.	Greaves, Joseph. ✱	1812 – 1816.
Blair, David.	1809 – 1814.	Grice, Joseph (II). ✱	1803 – 1804.
Blair, Jane Hannah.	1814 – 1819.	Hampton, Thomas. ✱	1812 – 1818.
Brown, Henry.	1805.	Harris, Joseph (II). ✱	1812 – 1818.
Clive & Son.	1816 – 1818.	Heely, James.	1815.
Clive & Turton. ✱	1811 – 1816.	Heely & Co., John. ✱	1816 – 1818.
Davies & Co., William.	1803 – 1808.	Heely & Co., John & Samuel. ✱	1803 – 1815.
Dawes, John.	1816 – 1819.	James, Henry.	1815 – 1816.
Dawes, Samuel & John.	1813 – 1816.	Ketland, William.	1804.
Dawes, William & Samuel.	1803 – 1813.	Marwood, William. ✱	1803 – 1810.

Morris, Henry.	1804 – 1810.		Dick, Walter. ✱	1803 – 1811.
Morris & Grice. ✱	1810 – 1819.		Duce, John.	1806 – 1818.
Moxham, Thomas. ✱	1810 – 1819.		Fletcher, Thomas.	1806 – 1818.
Muntz & Co., P.F. ✱	1812 – 1816.		Galton Jr., Samuel. ✱	1803 – 1819.
Osborn, Thomas. Barrel Rifler.	1806 – 1810.		Gill & Co. John. ✱	1805 – 1817.
Osborn & Gunby.	1808 – 1818.		Grice, Joseph (II). ✱	1803 – 1804.
Oughton, John. ✱	1803.		Ketland, Walker & Co. ✱	1808 – 1819.
Oughton, J.H.	1816.		Ketland & Allport.	1804 – 1819.
Parsons, Benjamin. ✱	1812 – 1818.		Ketland & Walker. ✱	1803 – 1808.
Parsons, Gad. ✱	1803 – 1813.		Leonard, William. ✱	1806 – 1818.
Parsons, John. ✱	1812 – 1818.		Marwood, William. ✱	1803 – 1810.
Parsons, Phineas. ✱	1813 – 1818.		Morris, Henry.	1804 – 1810.
Parsons & Son, Gad. ✱	1813 – 1818.		Moxham, Thomas. ✱	1810 – 1819.
Plant, Benjamin.	1812 – 1818.		Negus, James (II).	1806 – 1818.
Portlock, John.	1806 – 1809.		Osborn & Gunby.	1808 – 1818.
Portlock, Thomas. ✱	1803 – 1815.		Oughton, J.H.	1816.
Price, Theodore & Philomon.	1805 – 1815.		Peters, Michael. ✱	1806 – 1818.
Priest, Ann. ✱	1817 – 1818.		Rock, Joseph & James.	1816.
Priest, Joseph.	1805 – 1817.		Rock, Martha.	1810 – 1816.
Rock, Joseph & James.	1816.		Rock, William. ✱	1803 – 1810.
Rock, Martha.	1810 – 1816.		Round, Joseph. ✱	1803 – 1818.
Rock, William. ✱	1803 – 1810.		Shenton, William.	1806 – 1818.
Round, Benjamin. ✱	1812 – 1818.		Sherwood, Elizabeth & William.	1815.
Russell, John. ✱	1803 – 1818.		Sherwood, James.	1808.
Russell, Thomas. ✱	1803 – 1815.		Sherwood, John Jr. & James.	1807 – 1810.
Salter, John.	1808 – 1818.		Sherwood, Joseph Sr.	1803 – 1815.
Simes, John. ✱	1803 – 1805.		Sherwood, Joseph Jr. & James.	1805 – 1806.
Simes & Priest.	1804 – 1811.		Siddon, William.	1803 – 1813.
Stokes, Thomas. ✱	1803 – 1818.		Siddon & Sons.	1813 – 1818.
Sutherland, R. & R. ✱	1809 – 1818.		Smith, James. ✱	1803 – 1818.
Truman, Elizabeth. ✱	1803 – 1818.		Spittle, James.	1812.
Whately, Henry P.	1804 – 1815.		Spittle, Joseph.	1806 – 1818.
Whately, Henry & John.	1803 – 1816.		Spittle, Peter. ✱	1803 – 1818.
Whately, John. ✱	1803 – 1804.		Stone, Thomas. ✱	1806 – 1818.
Whately, John. ✱	1817 – 1819.		Sutherland, R. & R. ✱	1809 – 1818.
Wheeler, Robert Jr. ✱	1805 – 1808.		Whately, Henry & John.	1803 – 1816.
Wheeler & Son, Robert. ✱	1808 – 1819.		Whately, John. ✱	1803 – 1804.
Willetts & Holden. ✱	1805 – 1819.		Wheeler, Robert Jr. ✱	1803 – 1808.
			Wheeler & Son, Robert. ✱	1808 – 1819.
LOCKS			Whitehead, Francis. ✱	1803 – 1806.
Allport, William.	1816 – 1819.		Whitehead & Son, Francis.	1806 – 1818.
Bailey, James.	1808.		Wilkes, Job. ✱	1806 – 1818.
Bills, Samuel. ✱	1806 – 1818.		Willetts & Holden. ✱	1805 – 1819.
Blakemore, Thomas. ✱	1804 – 1818.		Woolley, Price & Jones.	1809 – 1816.
Blakemore, Thomas W.	1815.			
Blakemore & Robbins.	1807.			
Blakemore & Son, Thomas.	1807.			
Bulleis (Bulless), Thomas. ✱	1810 – 1816.			

RAMMERS, Steel.

Eames, Josiah.	1806 – 1816.
Galton Jr., Samuel.	1803 – 1819.
Grigg, Abigail.	1814 – 1816.
Grigg, John.	1812 – 1813.
Ketland, Elizabeth.	1816 – 1819.
Ketland & Allport.	1804 – 1819.
Morris & Grice. *	1810 – 1819.
Newman, Thomas.	1814 – 1816.
Osborn & Gunby.	1808 – 1818.
Oughton, J.H.	1816.
Rock, John.	1804 – 1816.
Rock, Martha.	1810 – 1816.
Rock, William. *	1803 – 1810.
Tranter, William.	1812 – 1818.
Whately, Henry & John.	1803 – 1816.
Whately, John. *	1803 – 1804.
Willmore, Joseph.	1813 – 1818.
Woolley, James. *	1803 – 1818.

BAYONETS

Allport, William.	1816 – 1819.
Bate, Thomas.	1807 – 1816.
Chambers, Samuel.	1812 – 1818
Cooper & Craven.	1803 – 1819.
Craven, Thomas.	1802 – 1803.
Dawes, John.	1816 – 1819.
Dawes, Samuel. *	1804 – 1805.
Dawes, Samuel & John.	1813 – 1816.
Dawes, William & Samuel. *	1804 – 1805.
Deakin, Francis.	1812 – 1813.
Gill, Elizabeth. *	1817 – 1819.
Gill & Co., John. *	1805 – 1817.
Hill, John. *	1812 – post 1840.
Ketland, Elizabeth. *	1816 – 1819.
Ketland & Allport.	1804 – 1819.
Makin, James. *	1803 – 1814.
Osborn, Henry. *	1803 – 1807.
Osborn & Gunby.	1808 – 1818.
Oughton, John. *	1803, 1812.
Oughton, John & Craxhall. *	1803 – 1815.
Oughton, J.H.	1816.
Reddell, Joseph H. *	1805 – 1809; 1815 – 1816.
Reddell & Bate.	1804 – 1807.
Rock, John.	1804 – 1816.
Rock, Joseph & James.	1816.
Rock, Joseph & Samuel. *	1816.
Rock, Martha.	1810 – 1816.
Rock, William. *	1803 – 1810.
Salter, George.	1812 – 1818.
Salter, John.	1808 – 1818.
Wheeler, Robert. *	1803 – 1808.
Wheeler & Son, Robert. *	1808 – 1819.
Woolley, James. *	1803 – 1818.
Woolley, Deakin, Dutton & Johnson.	1805 –
Woolley, Deakin & Dutton.	1809 –
Woolley & Co. *	1803 – 1819.
Woolley & Deakin.	1803 –

GUN FURNITURE

Crosbee, William.	1812 – 1814.
Harris, William.	1808 – 1811.
Mayor, Joseph. *	1804 – 1816; 1819.

IMPLEMENTS (Rifle buttbox)

Between April 1809 and the end of production in late 1814, Infantry Rifle box implements were supplied by the contractors with each rifle.

Sherwood, Joseph Sr. *	1806; 1814.

ROUGH STOCKERS & SETTERS UP

Nock, Henry. *	1803 – 1805.
Rea, John. *	1806 – 1808.
Reynolds, Thomas. *	1806, 1809.

COMPLETE ARMS

Archer & Son, Thomas.	1805 – 1818.
Allport, William.	1816 – 1819.
Ashton, Thomas.	1804 – 1818.
Barnett, Thomas. *	1804 – 1818.
Barnett & Son, Thomas.	1818 – 1819.
Brander, Martin. *	1803 – 1809.
Brander & Potts.	1809 – 1818.
Dawes, John.	1816 – 1819.
Dawes, Samuel & John. *	1813 – 1816.
Dawes, William & Samuel. *	1803 – 1813.
Egg, Durs. *	1803 – 1818.
Egg, Joseph. *	1816.
Fearnley, Ann. *	1813 – 1818.
Fearnley, Robert.	1804 – 1813.
Galton Jr., Samuel. *	1803 – 1819.
Gill, Elizabeth.	1817 – 1819.

Gill & Co., John. ✶	1805 – 1817.	Parr, James. Merchant.	1803 – 1810.
Grice, Joseph (II). ✶	1803 – 1804.	Patterson, Daniel.	1807 – 1809.
Hampton, Thomas. ✶	1812 – 1818.	Peake, John. ✶	1803 – 1818.
Hennem, Jonathan. ✶	1803 – 1805.	Pritchett, Richard Ellis. ✶	1809.
Hiscock, John.	1805.	Pritchett, Samuel. ✶	1803 – 1806.
Hollis, Richard & William. ✶	1812 – 1818.	Reynolds, Thomas. ✶	1806; 1809.
Ketland, Elizabeth. ✶	1816 – 1819.	Rolfe, Elizabeth.	1815 – 1816.
Ketland, William.	1804.	Rolfe, William Isaac.	1810 – 1815.
Ketland, Walker & Co. ✶	1808 – 1819.	Sutherland, R. & R. ✶	1809 – 1818.
Ketland & Allport.	1804 – 1819.	Thompson, James. ✶	1803 – 1808.
Ketland & Walker. ✶	1804 – 1808.	Thompson & Son, James. ✶	1808 – 1819.
Lowndes, Thomas.	1813 – 1818.	Whately, Henry & John.	1803 – 1816.
Marwood, Ann. ✶	1811 – 1818.	Whately, John. ✶	1803 – 1804.
Marwood, William. ✶	1803 – 1810.	Wheeler, Robert Jr. ✶	1803 – 1808.
Morris, Henry.	1804 – 1810.	Wheeler & Son, Robert. ✶	1808 – 1819.
Morris & Grice. ✶	1810 – 1819.	Willetts & Holden. ✶	1805 – 1819.
Moxham, Thomas. ✶	1810 – 1819.	Wilson & Son, William.	1803 – 1805.
New, Henry & Elizabeth.	1817 – 1818.	Wright & Co., Robert. ✶	1803 – 1819.
Parker, William. ✶	1804 – 1818.	Yeomans, James. ✶	1809.
Parkin, Thomas. Merchant.	1808 – 1810.		

1820 – 1839

With the exception of some Land Service Pistols in 1821, all components supplied and arms production between 1820 and 1827 relate to Infantry Rifles. The production of (Pattern 1823) rifles is the first instance of the recently established Royal Small Arms Factory at Enfield Lock being used to produce a large number (5,000) of newly made arms. There was clearly a strong effort made to make Enfield self-sufficient in the matter of components, and barrels, locks and brasswork were indeed fabricated at the new establishment, but not in such quantities as to entirely eliminate the private sector trade. Due to a lack of records for the RSAF in its early years, it is not possible to say whether the small groups of arms produced by the contractors throughout this period were intended largely as price-control exercises performed concurrently with similar work at Enfield, or simply that these small orders represent the minimal needs of the service.

During the short reign of William IV (1830 - 1837) serious experimentation with the percussion ignition system for military small arms was carried out and ultimately adopted. Several varieties of carbine were produced, including the Pattern 1835 'Manton' carbine, some 16 inch barrelled carbines for the forces of Queen Isabella II of Spain (1835 - 1836), a small number of Eliott carbines and a small number of Land and Sea Service pistols in 1836 and 1837. Pattern 1823 Infantry Rifles for the Shah of Persia were produced in 1836, and an additional number for Government use in 1838 and 1839. Locks produced after June 1837 will bear the crowned cypher of Queen Victoria.

BARRELS (Infantry Rifle)

Adams, John.	1825 – 1826.
Bayliss & Son, Joseph.*	1824.
Beckwith, William A.	1838. See entry.
Blair, Jane Hannah.*	1823.
Brander & Potts.*	1820; 1824 – 1827.
Bulleis, Thomas.*	1823.
Clive, John.	1825; 1838.
Clive, John Jr.	1824.
Clive, Thomas.	1823.
Clive & Turton.*	1823 – 1825.
Dawes, Samuel & John.*	1823.
Deakin, Mary.	1823 – 1824.
Deakin, Samuel.	1823.
Deakin, William.	1823.
Deakin & Son, Francis.*	1823.
Gill & Sons, Elizabeth.	1823.
Greaves, Joseph.*	1823 – 1824.
Gunby, John.	1823.
Harris, Joseph (II).*	1823 – 1824.
Heely & Co., John.*	1823.
Johnson, Francis.	1823.
Meredith & Moxham.	1823.
Muntz & Co., P.F.*	1823, 1825.
Osborn, Henry.*	1823; 1825; 1826.
Oughton, John & Craxhall.*	1823 – 1826.
Parsons, Benjamin.*	1823; 1825.
Parsons, John.*	1823.
Parsons, Phineas.*	1823.
Parsons & Son, Gad.*	1823.
Plant, Martha.	1823.
Priest, Ann.*	1823.
Reddell, Joseph H.*	1823.
Rock, Joseph & Samuel.*	1823 – 1824.
Round, Benjamin.*	1823 – 1824.
Russell, John.*	1823.
Stokes, Thomas.*	1823.
Truman, Elizabeth.*	1823 – 1824.
Turner, Joseph.	1823; 1838.
Whately, John.*	1823.
Wheeler, John.	1823.
Wheeler, Robert.	1823.
Willetts & Holden.*	1823 – 1824.
Woolley & Sargeant.	1823.

LOCKS (Infantry Rifle)

Allport, William.*	1823.
Ashmore, R.	1838.
Ashton, J.	1838.
Bayliss, Lewis.	1824.
Bills, Richard.	1823 – 1826.
Bills, Samuel.*	1826.
Blakemore, J.P.	1823 – 1826
Blakemore, Mary.	1823 – 1826.
Bulleis, Thomas.*	1823.
Corbett, J.	1838.
Deakin, Mary.	1823 – 1824.
Deakin, Samuel.	1823.
Deakin, William.	1823.
Deakin & Son, Francis.	1823.
Duce, J.	1838.
Duce, Mary & John.	1825 – 1826.
Fletcher & Co.	1823.
Gill & Sons, Elizabeth.	1823.
Gunby, John.	1823.
Howell, William.	1823 – 1825.
Johnson, Francis.	1823.
Ketland, Elizabeth.*	1823.
Ketland, Thomas.	1823.
Ketland, Walker & Co.*	1823.
Leonard, William.*	1823.
Negus, Deborah.	1823.
Osborn, Henry.*	1823; 1825; 1826.
Oughton, John & Craxhall.*	1823 – 1826.
Partridge, John.	1823 – 1826.
Partridge, W.	1838.
Peters, Michael.*	1823.
Price, Theodore.	1823; 1826.
Reddell, Joseph H.	1823.
Rock, Samuel.	1823.
Rubery, J.	1838.
Sansom, S.	1838.
Shenstone, Hannah.	1823.
Sherwood, J. & W.	1823 – 1826.
Siddon & Co.*	1823.
Slater, John.	1823 – 1826.
Smith, James.*	1823 – 1826.
Spittle, M. & F.	1823; 1824.
Spittle, Peter.*	1823.
Stone, Thomas.*	1823 – 1826.
Sutherland, Ramsay & Richard.*	1823.
Thornhill, Charles.	1823 – 1826.

Turner, B.	1838.	Lacy & Reynolds.	1836.
Wheeler & Son, Robert.✶	1823, 1826.	Leigh, James Brooks.	1838.
Whitehead, J.	1838.	Marwood, Ann.✶	1820; 1824; 1825.
Whitehead & Co.	1823 – 1826.	Mills, William.	1839.
Wilkes, Job.✶	1823 – 1826.	Mills & Son, William.	1836; 1837.
Willetts & Holden.✶	1823.	Morris & Grice.✶	1823.
Wilson, C.	1838.	Moxham, Thomas.✶	1825.
Woolley & Price.	1823.	Parker, William.✶	1820 – 1821; 1824 – 1827; 1836 – 1839.
Woolley & Sargeant.	1823.	Peake, John.	1820 – 1821.
Yates, J.	1838.	Potts, Thomas.	1836 – 1839.

BAYONETS (Infantry Rifle)
Mole, John & Robert.   1836; 1838.

Pritchett, Richard Ellis.✶   1824 – 1827; 1835 – 1839.

GUN FURNITURE (Infantry Rifle)
Glascott, Mary & G.M.	1823 – 1826.
Glascott Brothers.	1838.

Reynolds, Thomas.✶   1825, 1827.
Reynolds & Son, Thomas.   1836 – 1838.
Sargant & Son.   1836 – 1837.
Sutherland, Ramsay & Richard.✶ 1825.
Thomas. William.   1838; 1839.
Thompson & Son, James.✶   1821; 1824; 1825; 1827; 1836; 1837.
Wheeler & Son, Robert.✶   1825.
Wright, Robert.✶ 1820; 1824; 1825; 1827.
Yeomans, James.✶   1836.
Yeomans & Son, James.   1835 – 1839.

IMPLEMENTS (Rifle buttbox)
Aston, Joseph.	1838.
Baker & Son, E.	1823; 1825; 1826; 1827.
Sherwood, Joseph.	1823; 1825; 1826.
Wheeler & Son, Robert.✶	1835.

ROUGH STOCKED & SET UP
Ashton, T. & C.	1824 – 1827.
Ashton, Thomas.	1836 – 1839.
Baker, Ezekiel.✶	1820 – 1825.
Baker, Ezekiel John.	1836 – 1839.
Baker & Son, E.	1826
Barnett, John Edward.	1832 – 1839.
Barnett & Son, Thomas.✶	1820 – 1827.
Blair, Jane Hannah.	1823.
Bond, William James.	1836.
Bond, William Thomas.	1835; 1837.
Egg, Durs.✶	1821 – 1827.
Egg, Joseph.✶	1820; 1824 – 1827; 1836.
Fearnley, Ann.	1820; 1824.
Hampton, Thomas.✶	1825.
Heptinstall, William.	1838; 1839.
Hollis, Richard & William.✶	1825.
Ketland & Co., W.	1825.
Lacy & Co.	1838.

FINIS

NOTES

NOTES